Atanu Sengupta and Tapan Kumar Pal

Fuzzy Preference Ordering of Interval Numbers in Decision Problems

Studies in Fuzziness and Soft Computing, Volume 238

Editor-in-Chief

Prof. Janusz Kacprzyk
Systems Research Institute
Polish Academy of Sciences
ul. Newelska 6
01-447 Warsaw
Poland
E-mail: kacprzyk@ibspan.waw.pl

Further volumes of this series can be found on our homepage: springer.com

Vol. 223. Oscar Castillo, Patricia Melin
Type-2 Fuzzy Logic: Theory and Applications, 2008
ISBN 978-3-540-76283-6

Vol. 224. Rafael Bello, Rafael Falcón, Witold Pedrycz, Janusz Kacprzyk (Eds.)
Contributions to Fuzzy and Rough Sets Theories and Their Applications, 2008
ISBN 978-3-540-76972-9

Vol. 225. Terry D. Clark, Jennifer M. Larson, John N. Mordeson, Joshua D. Potter, Mark J. Wierman
Applying Fuzzy Mathematics to Formal Models in Comparative Politics, 2008
ISBN 978-3-540-77460-0

Vol. 226. Bhanu Prasad (Ed.)
Soft Computing Applications in Industry, 2008
ISBN 978-3-540-77464-8

Vol. 227. Eugene Roventa, Tiberiu Spircu
Management of Knowledge Imperfection in Building Intelligent Systems, 2008
ISBN 978-3-540-77462-4

Vol. 228. Adam Kasperski
Discrete Optimization with Interval Data, 2008
ISBN 978-3-540-78483-8

Vol. 229. Sadaaki Miyamoto, Hidetomo Ichihashi, Katsuhiro Honda
Algorithms for Fuzzy Clustering, 2008
ISBN 978-3-540-78736-5

Vol. 230. Bhanu Prasad (Ed.)
Soft Computing Applications in Business, 2008
ISBN 978-3-540-79004-4

Vol. 231. Michal Baczynski, Balasubramaniam Jayaram
Soft Fuzzy Implications, 2008
ISBN 978-3-540-69080-1

Vol. 232. Eduardo Massad, Neli Regina Siqueira Ortega, Laécio Carvalho de Barros, Claudio José Struchiner
Fuzzy Logic in Action: Applications in Epidemiology and Beyond, 2008
ISBN 978-3-540-69092-4

Vol. 233. Cengiz Kahraman (Ed.)
Fuzzy Engineering Economics with Applications, 2008
ISBN 978-3-540-70809-4

Vol. 234. Eyal Kolman, Michael Margaliot
Knowledge-Based Neurocomputing: A Fuzzy Logic Approach, 2009
ISBN 978-3-540-88076-9

Vol. 235. Kofi Kissi Dompere
Fuzzy Rationality, 2009
ISBN 978-3-540-88082-0

Vol. 236. Kofi Kissi Dompere
Epistemic Foundations of Fuzziness, 2009
ISBN 978-3-540-88084-4

Vol. 237. Kofi Kissi Dompere
Fuzziness and Approximate Reasoning, 2009
ISBN 978-3-540-88086-8

Vol. 238. Atanu Sengupta, Tapan Kumar Pal
Fuzzy Preference Ordering of Interval Numbers in Decision Problems, 2009
ISBN 978-3-540-89914-3

Atanu Sengupta and Tapan Kumar Pal

Fuzzy Preference Ordering of Interval Numbers in Decision Problems

Springer

Authors

Dr. Atanu Sengupta
Formerly with,
Department of Applied Mathematics with
Oceanology & Computer Programming
Vidyasagar University
Medinipur - 721 102, India
Present Address:
2BR-F4, I.I.T. Kharagpur Campus,
Technology Kharagpur – 721302, India
E-Mail: atanu_sgupta@yahoo.co.in

Prof. Tapan Kumar Pal
Department of Applied Mathematics with
Oceanology & Computer Programming
Vidyasagar University
Medinipur - 721 102
India

ISBN 978-3-540-89914-3 e-ISBN 978-3-540-89915-0

DOI 10.1007/978-3-540-89915-0

Studies in Fuzziness and Soft Computing ISSN 1434-9922

Library of Congress Control Number: 2008941013

Typeset & Cover Design: Scientific Publishing Services Pvt. Ltd., Chennai, India.

Printed in acid-free paper

9 8 7 6 5 4 3 2 1

springer.com

To Proma

Preface

In conventional mathematical programming, coefficients of problems are usually determined by the experts as crisp values in terms of classical mathematical reasoning. But in reality, in an imprecise and uncertain environment, it will be utmost unrealistic to assume that the knowledge and representation of an expert can come in a precise way.

The wider objective of the book is to study different real decision situations where problems are defined in inexact environment. Inexactness are mainly generated in two ways – (1) due to imprecise perception and knowledge of the human expert followed by vague representation of knowledge as a DM; (2) due to hugeness and complexity of relations and data structure in the definition of the problem situation. We use interval numbers to specify inexact or imprecise or uncertain data. Consequently, the study of a decision problem requires answering the following initial questions –

How should we

- compare and define preference ordering between two intervals?
- interpret and deal inequality relations involving interval coefficients?
- interpret and make way towards the goal of the decision problem?

The present research work consists of two closely related fields: first-one approaches towards defining a generalized preference ordering scheme for interval attributes. We have suggested two indices, *viz.*, Acceptability Index and Fuzzy Preference Ordering for interval attributes. But, before that a detailed comparative study on the existing ordering indices was necessary to identify the lacunae. Here we liked to incorporate possibly all relevant and even the most recent literatures up to 2008. Naturally, this part of our contribution, spanned more or less within Chapter 1 – 4, has taken considerable room in this book.

The next related field approaches to deal with some issues having application potential in many areas of decision making. Here we tried to develop mathematical description of relationships using interval arithmetic or re-structure some optimization models in inexact environment using interval numbers. Development of working algorithm for these models has also been attempted.

This book consists of nine chapters including the introductory first chapter, which gives a brief review of the literature and basic intention of this study. Our research contribution constitutes Chapter 2 to Chapter 8. The last chapter gives the chapter-wise summary and indicates the direction and scope in brief of the future research.

It is a great opportunity to this publication under the series 'Studies in Fuzziness and Soft Computing' by Physica-Verlag (A Springer-Verlag company). We wish to thank the publishers.

We sincerely thank Professor Ajoy Kumar Ray (Head, SMST & Professor, E&ECE Dept., I.I.T. Kharagpur) for his enthusiastic encouragement and support in the preparation of this manuscript. We are fortunate to have constant support and invaluable suggestions from Dr. Debjani Chakraborty (Dept. of Maths., I.I.T. Kharagpur) in all phases of our work. We deeply thank her. We are greatly indebted to Professor Debasish Mondal (Dept. of Economics, VU) for his sincere inspiration and support. We are indebted to Professor R.N. Jana and Professor M.M. Pal of our department for their cordial support in many ways. We also like to extend our sincere thanks to all of our friends and colleagues who supported us in numerous ways in preparing this book.

Atanu Sengupta
Tapan Kumar Pal

Contents

Chapter 1

Introduction

"There must be an ideal world, a sort of mathematician's paradise where everything happens as it does in textbooks."

-- Bertrand Russell (1872-1970)

In our quest for such an ideal world with utmost precision and certainty we often try to replicate the pervasiveness of the real world into a formal model that aims to be a kind of abstraction of domain and the forces and dynamics of the environment.

But in reality, formal (mathematical) models happen to be based on precise information, certain assumptions, crisp hypotheses and well-defined theories, which are logically sound but very little, can mimic reality. For example, formal (mathematical) models used in decision aid are usually characterized by multiple parameters. A situation with *precise information* occurs when the Decision Makers (DMs) are able to indicate a precise value for each parameter. However, there are always many difficulties for obtaining precise values for all the parameters:

- the performance of each action on each criterion may be unknown at the time of the analysis – it may result from arbitrariness in constructing parts of the model or from the aggregation of several aspects having impact on different criteria, and it may result from a measuring instrument or from a statistical tool (which usually involves certain or uncertain extent of error);

A. Sengupta and T.K. Pal: Fuzzy Preference Ordering, STUDFUZZ 238, pp. 1–23.
springerlink.com

- not all but many parameters, which define relations between the effects and consequences while framing parts of the model, generally have seldom objective existence – they are mostly reflected by the DMs' subjective perception and opinion, which the DM may find difficult to express in so-called 'mathematical precision' and an objective measure of which may change from man-to-man in time-to-time;
- in either case, the result of a quest for numerical precision and certainty is neither real nor accurate. A decision situation related to human aspect, in fact, has only a little to do with the absolute attributes – certainty and precision – which are not present in our cognition, perception, reasoning and thinking. There are so many things which can only be defined by vague and ambiguous predicates and as a result it has been increasingly clear that formal modelling of a real decision situation does not reflect the pervasiveness of human perception, cognition and mutual interaction with the outside world (Gupta (1988)).

I would like to quote from the works of Sir Karl Popper (1974) on the philosophy of science in this context:

> Both precision and certainty are false ideals. They are impossible to attain, and therefore dangerously misleading if they are uncritically accepted as guides. The quest for precision is analogous to the quest for certainty, and both should be abandoned... one should never try to be more precise than the problem situation demands…

Quite a similar note can be found in the words of one of the greatest scientists of all times, Albert Einstein (1879-1955), who says:

> "As far as the laws of mathematics refer to reality, they are not certain; and as far as they are certain, they do not refer to reality."

As a matter of fact, in quest of realistic model building and analyses, the representation and manipulation of these *inexact*, *vague*, *fuzzy*, *ambiguous* or *imprecise* information and relations that abound in real world situations become one of the major concerns to the *new-era mathematicians* (the proponents of *fuzzy-ism*, after Zadeh (1962, 1965)'s pioneering call) and since the late 1980's this ensued in an highly demanding field of research, viz., *computing with words*. Different researchers used different distinct expressions to name this type of information:

- *imprecise information* is used in the works of Athanassopoulos & Podinovski (1997), Miettinen & Salminen (1999),
- the term *incomplete information* is used by Weber (1987),

- *partial information* is used by Hazen (1986) and also by Deng (1982) to define his *grey systems theory*.
- *Poor information* is used by Bana e Costa & Vincke (1995).
- *nonpoint information*, a new designation of such information is given by Dias (web document).

This concept of *imprecise information* generalizes the concepts of multiple states of affairs and the concepts of interval parameters. At a given stage of the decision process, there may be a set of the multiple acceptable combinations of parameter values or may be a set of combinations compatible with the available imprecise information (Dias (web document)). This data set may be discrete or continuous. In the latter case the set is defined by the bounds for the parameter values what we commonly call as numerical interval information.

1.1. Conventional distinction between uncertainty and imprecision

Prior to Zadeh's pioneering work on fuzzy sets, the probability theory based on Boolean Logic was an important tool to deal with uncertainties (or randomness) of real events and activities. Fuzzy set theory provided a valuable conceptual tool for dealing with non-stochastic imprecision or so-called vague concepts. Both the notions cover two different types of ignorance and are dealt with two different types of theories with ideas of their own.

Probabilistic uncertainty relates to events that have a well-defined, unambiguous meaning. In probability theory, the notion of probability is required only to quantify our ignorance (uncertainty) about which particular case is going to be observed. We perfectly know the set and the distribution of the elements (in terms of the properties of the underlying population) in the set but we do not know which particular case will be observed in a random experiment. The notion of probability is used to express a forecast about belongingness or not belongingness of an event (that is going to be observed) in a set that is perfectly well known. Probability theory is based on classical set theory and on two-valued logic, e.g. true-or-false or yes-or-no statements; probability theory assesses whether an event will occur (Klir & Folger (1988), Kosko (1990)).In contrast, the fuzzy set theory is based on the notion that the individual elements of a set are well known but the set is not well-defined or ill-defined in such a way that one simply can not decide by 'yes' or 'no' about the belongingness of an element in the particular set. Fuzzy set theory is based on multi-valued logic (Zimmermann (1991), McNeill & Freiberger (1993), Klir & Yuan (1995), Chakraborty (1995)).

The probabilities represent uncertainties but not the degrees of partial truths, which are used for the description of imprecision by degrees of membership. If imprecision is the state of nature of a situation and the resulting uncertainty is possibilistic rather than probabilistic, then the situation is said to be fuzzy (Buckley (1987), (Babad & Berliner (1994)). Probability theory and fuzzy set theory are obviously not the alternative concepts, but are well accepted complementary tools to describe many specific instances of uncertainty and imprecision (Chakraborty (1995), Mukaidono (2001)). Zadeh (1978) has used the term *vagueness* to designate *fuzziness* and *ambiguity*, however, Dubois & Prade (1980)'s *fuzziness* is *vagueness* but it differs from *ambiguity* and *generality*.

1.2. Imprecise data representation: Preliminaries on Fuzzy set and genesis of Interval Numbers as imprecise data

The concept of a fuzzy set was introduced by Zadeh (1965) to represent or manipulate data and information possessing non-statistical uncertainties.

In classical set theory, a subset A of a set X can be defined by its characteristic function χ_A as a mapping from the elements of X to the elements of the set $\{0, 1\}$,

$$\chi_A : X \to \{0, 1\}.$$

This mapping may be represented as a set of ordered pairs, with exactly one ordered pair for each element of X. The truth or falsity of the statement "x is in A" is determined by the ordered pair $(x, \ \chi_A(x))$. The statement is true if $\chi_A = 1$ and the statement is false if $\chi_A = 0$.

Similarly, a fuzzy subset $\tilde{A}$ in a universe of discourse X is characterized by its membership function $\mu_{\tilde{A}}$ as a mapping from the elements of X to the values of the interval $[0, 1]$, such that,

$$\mu_{\tilde{A}} : X \to [0, 1].$$

The degree to which the statement "x is in fuzzy set $\tilde{A}$" is true is denoted by the set of ordered pairs $(x, \ \mu_{\tilde{A}}(x))$, where $\mu_{\tilde{A}}(x)$ is interpreted as the degree of membership of element x in fuzzy set $\tilde{A}$. A fuzzy set defines the concept of a set with unsharp (imprecise or fuzzy) boundary between the membership and non-membership of its elements (Kaufmann & Gupta (1991)).

Definition: The *support* of the fuzzy subset $\tilde{A}$, denoted as, $\text{supp}(\tilde{A})$ is the crisp subset of X having all elements with non-zero membership grades in $\tilde{A}$. Notationally stated,

$$\text{supp}(\tilde{A}) = \left\{ x \in X \,\middle|\, \mu_{\tilde{A}}(x) > 0 \right\}.$$

Synonyms of support can be the degree/extent of fuzziness or the fuzzy spread (Prodanovic & Simonovic (2002)).

Definition: The height of a fuzzy set is the largest membership grade attained by any element in that set. A fuzzy set $\tilde{A}$ in the universe of discourse X is called *normalized* when the height of $\tilde{A}$ is equal to 1 (Klir & Yuan (1995)).

Definition: A fuzzy set $\tilde{A}$ in the universe of discourse X is convex, if and only if

$$\mu_{\tilde{A}}(\lambda x_1 + (1-\lambda)x_2) \;\; \geq \;\; \min(\mu_{\tilde{A}}(x_1), \mu_{\tilde{A}}(x_2))$$

for all x_1, x_2 in X and all $\lambda \in [0, 1]$, where *min* denotes the minimum operator (Klir & Yuan (1995)).

Definition: The α-level set (or α-cut) of a fuzzy set $\tilde{A}$ of X is a non-fizzy set (or a crisp interval) denoted by $[\tilde{A}]^{\alpha}$ and defined as

$$[\tilde{A}]^{\alpha} = \left\{ x : \mu_{\tilde{A}}(x) \geq \alpha, \;\; x \in X \right\}, \text{if } \alpha \in (0, 1]$$

$$= \text{cl}(\text{supp}(\tilde{A})), \;\text{if } \alpha = 0,$$

where, $\text{cl}(\text{supp}(\tilde{A}))$ denotes the closer of support of $\tilde{A}$ (Dubois & Prade (1980)).

Definition: A *fuzzy number* (or more generally, *fuzzy quantity*) $\tilde{N}$ is a convex and normalized fuzzy subset of the real line $\Re$ (Dubios & Prade (1980), Kaufmann & Gupta (1991)).

In many situations we often summarize numeric information, as for example, around Rs. 5000, near zero, about $10^{\circ}C$, about 15 – 20%, possibly not less than 2000 units. These sorts of numerically transmittable data are not precise or crisp in terms of classical mathematical reasoning but are very meaningful in terms of human communication, perception and reasoning. These imprecise data could be the examples of what are called *fuzzy numbers*.

The membership function of a fuzzy number $\tilde{N}$ has the following properties:

(i) $\mu_{\tilde{N}}(x) = 0$, outside of some interval $[a, d]$;

(ii) There are real numbers b and c, $a \le b \le c \le d$ such that $\mu_{\tilde{N}}(x)$ is monotone increasing on the interval $[a, b]$ and monotone decreasing on the interval $[c, d]$;

(iii) $\mu_{\tilde{N}}(x) = 1$, for each $x \in [b, c]$.

If $\tilde{M}$ is a fuzzy number, then $[\tilde{M}]^{\gamma}$ is a closed interval of $\Re$ for all $\gamma \in [0, 1]$. Here we introduce an alternative notation of $[\tilde{M}]^{\gamma}$ as,

$$[\tilde{M}]^{\gamma} = [m_1(\gamma), m_2(\gamma)] \in \Re .$$

where, $m_1(\gamma)$ and $m_2(\gamma)$ are the *lower* and *upper bounds* of the interval $[\tilde{M}]^{\gamma}$.

A fuzzy set $\tilde{A}$ is called a *triangular fuzzy number* with peak (or center) a, left width $\alpha \ge 0$ and right width $\beta \ge 0$ and is denoted as $\tilde{A} = (a, \alpha, \beta)$, if its membership function has the following form,

$$\mu_{\tilde{A}}(x) = \begin{cases} \dfrac{x - (a - \alpha)}{\alpha} & \text{if} \quad a - \alpha \le x \le a, \\ \dfrac{(a + \beta) - x}{\beta} & \text{if} \quad a \le x \le a + \beta, \\ 0 & \text{otherwise.} \end{cases}$$

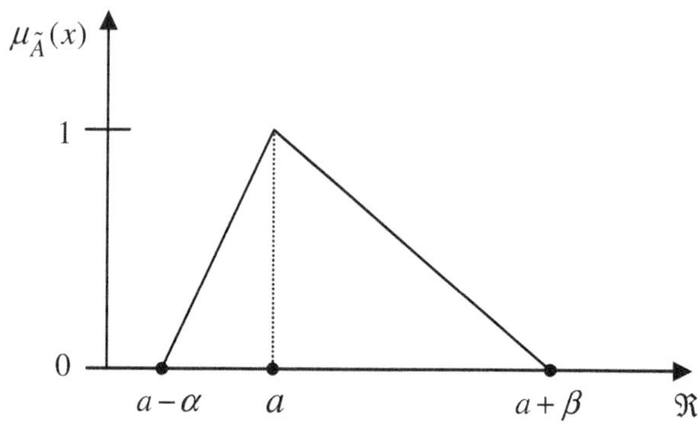

Fig. 1.1: Triangular fuzzy number.

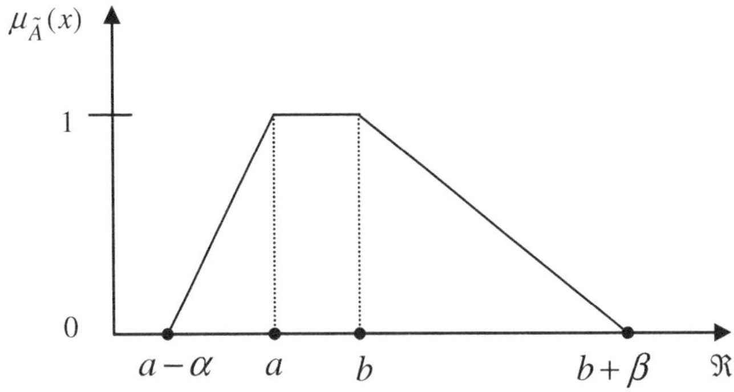

Fig. 1.2: Trapezoidal fuzzy number.

Now, using γ -cut of $\tilde{A}=(a,\ \alpha,\ \beta)$, it can easily be verified that

$$[\tilde{A}]^{\gamma}=[a-(1-\gamma)\alpha,\ a+(1-\gamma)\beta],\quad \forall\gamma\in[0,1].$$

Fig. 1.1 shows a triangular fuzzy number with center *a*, which can be cited as an example of a fuzzy data or fuzzy quantity, '*x* is approximately equal to *a*'.

A fuzzy set $\tilde{A}$ is called a *trapezoidal fuzzy number* with tolerance interval $[a,b]$, $a\le b$, left width $\alpha\ge 0$ and right width $\beta\ge 0$ and is denoted as $\tilde{A}=(a,\ b,\ \alpha,\ \beta)$, if its membership function has the following form

$$\mu_{\tilde{A}}(x)=\begin{cases}\dfrac{x-(a-\alpha)}{\alpha} & \text{if } a-\alpha\le x\le a,\\ 1 & \text{if } a\le x\le b,\\ \dfrac{(b+\beta)-x}{\beta} & \text{if } a\le x\le b+\beta,\\ 0 & \text{otherwise.}\end{cases}$$

In Fig. 1.2, a trapezoidal fuzzy number is shown which represents the imprecise statement, "*x* is approximately in the interval [*a*, *b*]". A trapezoidal fuzzy number is often called a *flat fuzzy number* or a *fuzzy interval* or *LR-type fuzzy number* (Zadeh (1975), Dubois & Prade (1980)). The trapezoidal fuzzy number reduces to a triangular fuzzy number if $a=b$ and it reduces to an R-fuzzy number, if $\alpha=0$ and an L-fuzzy number, if $\beta=0$.

Similarly, the γ-cut of $\tilde{A}=(a,\ b,\ \alpha,\ \beta)$ can easily be verified that

$$[\tilde{A}]^{\gamma}=[a-(1-\gamma)\alpha,\ b+(1-\gamma)\beta],\quad \forall\gamma\in\ [0,1].$$

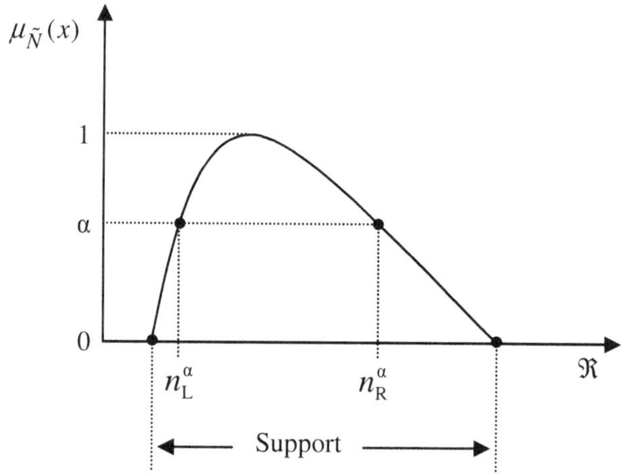

Fig.1.3: The α-cut of a fuzzy number.

Therefore, for a fuzzy set $\tilde{N}$ in *X*, the symbol $[\tilde{N}]^\alpha$ represents a non-empty non-fuzzy bounded interval contained in *X* and can be denoted by

$$[\tilde{N}]^\alpha = [n_L^\alpha,\ n_R^\alpha] = \left\{ t \,\middle|\, n_L^\alpha \le t \le n_R^\alpha \right\},$$

where, n_L^α and n_R^α are respectively the lower and upper bounds of the closed interval (Kaufmann & Gupta (1991), Zimmermann (1991)). The essence of the α-cut is that it limits the domain under consideration to the set of elements with the degree of membership of at least alpha. Thus, while the support (extent of fuzziness) of fuzzy set $\tilde{N}$ is its entire base, its α-cut is from n_L^α to n_R^α. Values outside the interval $[n_L^\alpha,\ n_R^\alpha]$ are considered to have a level of membership too insignificant to be relevant and should be excluded (cut out) from consideration. Synonyms of support are degree of fuzziness or a fuzzy spread.

Three fundamental operations in classical set theory are union, intersection and complement. Since membership in a fuzzy set is a matter of degree, set operations are generalized accordingly. Union, intersection and complement operations in fuzzy set theory are similar to disjunction ('*logical or*'), conjunction ('*logical and*') and negation from a logical point of view (Zimmermann (1991), Yen & Langari (1999)).

A common fuzzy disjunction operator is the *maximum* operator and hence, *fuzzy union* of two fuzzy sets $\tilde{A}$ and $\tilde{B}$ (Fig. 1.4) is mostly defined as

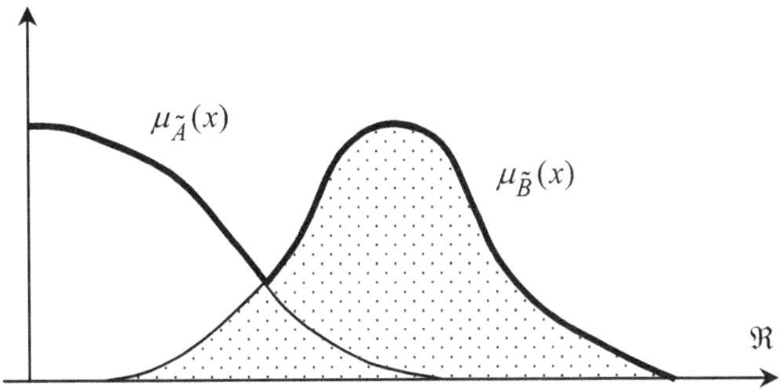

Fig. 1.4: Fuzzy union: the membership function $\mu_{\tilde{A}\cup\tilde{B}}$ is shown by the bold line.

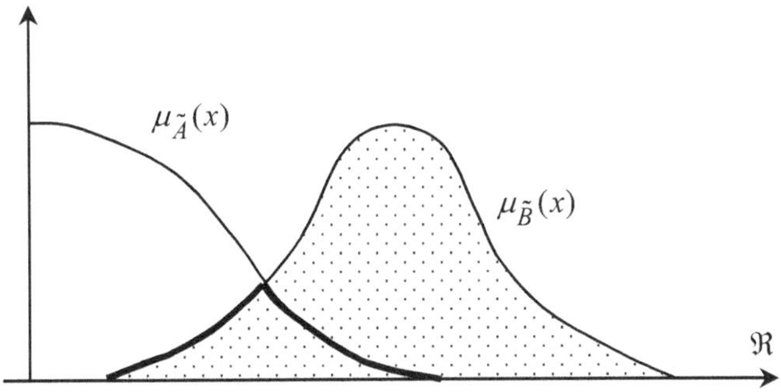

Fig. 1.5: Fuzzy intersection: the membership function $\mu_{\tilde{A}\cap\tilde{B}}$ is shown by the bold line.

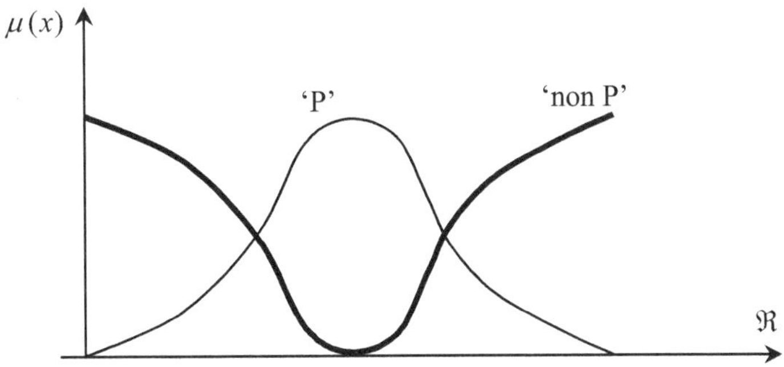

Fig. 1.6: Fuzzy premise 'P' and its complement.

$$\mu_{\tilde{A}\cup\tilde{B}}(x)=\max\left\{\mu_{\tilde{A}}(x),\mu_{\tilde{B}}(x)\right\}.$$

Similarly, a common fuzzy conjunction operator is the *minimum* operator and so *fuzzy intersection* of two fuzzy sets $\tilde{A}$ and $\tilde{B}$ (Fig. 1.5) is defined as

$$\mu_{\tilde{A}\cap\tilde{B}}(x)=\min\left\{\mu_{\tilde{A}}(x),\mu_{\tilde{B}}(x)\right\}.$$

The membership function of the *complement* of a fuzzy set $\tilde{A}$ is defined as

$$\mu_{\tilde{A}^{c}}(x)=1-\mu_{\tilde{A}}(x)$$

where $\tilde{A}^{\mathrm{c}}$ is complement of $\tilde{A}$. Therefore, if 'P' is true, its complement 'non P' is false, or if $\mu_{\mathrm{P}}=1$ then $\mu_{\mathrm{non\ P}}=0$ and vice versa. (Fig. 1.6).

1.3. Interval Numbers: A better tool to represent imprecision and uncertainty

In order to develop a good Decision Science methodology, fuzzy and stochastic approaches are frequently used to describe and treat imprecise and uncertain elements present in a decision problem.

The probabilistic analysis is the most widely used method for characterizing uncertainty in physical systems, especially when estimates of input probability distributions are available. It can describe uncertainty arising from stochastic disturbances, variability conditions, and risk considerations. In literature, in stochastic optimization (Charnes & Cooper (1959), Vajda (1972), Sengupta (1981), Kall (1982), Santoso et al (2005), Talluri et al (2006)), parameters are assumed to have known probability distributions.

Similarly, fuzzy theory facilitates uncertainty analysis of systems where uncertainty is due to *vagueness* or *fuzziness* rather than due to randomness alone and can be applied to uncertainty analysis with imprecise observation values or with linguistic data or verbal descriptions of relations. However, the membership function, the core of any fuzzy model, is considered to be both the strongest and the weakest point of fuzzy set theory (Munda et al (1992), Cornelissen et al (2000)). The strongest point is that it defines a soft threshold, in contrast with a characteristic function defining a hard threshold in classical set theory. It is the weakest, because the membership function is regarded as too subjective in relation to its construction. In industrial engineering applications of fuzzy set theory, construction of membership functions is realized mostly by trial and error (Zimmermann (1991), McNeill & Freiberger (1993), Bárdossy & Duckstein (1995), Klir& Yuan (1995)).

Several studies discuss empirical methods to construct a membership function based on expert knowledge (e.g., Norwich & Türksen (1984), Chameau & Santamarina (1987), Santamarina & Chameau (1987), Türksen (1991, 1999), Bárdossy & Duckstein (1995)). In the works of fuzzy optimization, fuzzy parameters are assumed to be with known membership functions (e.g., Luhandjula (1983, 1989), Slowinski (1986), Delgado et al (1989), Rommelfanger (1989), Chen (2001)).

As a matter of fact, to a DM it is not always easy to specify the membership function or probability distribution in an inexact environment. In reality, while building any decision methodology, it seems in principle contradictory to analyze the parametric *vagueness* and *lack of information* by separate models and assumptions (in order to shape the membership function or the probability distribution), which again require considerable amount of additional and reliable knowledge.

Apart from the fuzzy theory and probabilistic analysis, the existing methods for characterization of imprecision and uncertainty include another way, the use of interval mathematics. At least in some of the cases, the use of interval coefficients may serve the purpose better (Alefeld & Mayer (1996)). An interval number is the simplest and best representation of *imprecision* and *uncertainty* where no assumptions are made as to *shape* the uncertainty between lower and upper bound of an evaluation of any economic or technological consequence. All sorts of uncertainties can be transformed into one interval having only the bounds of the parameter concerned (Schjaer-Jacobsen (1996, 2002), Kearfott & Kreinovich (1996), Taguchi et al (1998), Vehí et al (2002)).

An interval number can be thought as

(a) an extension of the concept of a real number and also as a subset of the real line $\Re$ (Moore (1979), Alefeld & Herzberger (1983)).

(b) α–cut of a convex fuzzy set (Kaufmann & Gupta (1985), Lodwick & Jamison (2003)).

(c) a degenerated flat fuzzy number or fuzzy interval with zero left and right spreads, or, in other words, as the simplest fuzzy number (Schjaer-Jacobsen (2002)).

(d) a simplest form tolerance-type uncertainty with no information about the probabilities within this tolerance range (Nguyen & Kreinovich (2005)).

(e) a *grey number* whose exact value is unknown but a range within which the value lies is known ((Huang et al. (1995), Liu & Lin (1998), Yang & John (2003), Karmakar & Mujumdar (2006)).

As a coefficient, an interval signifies the extent of tolerance or a region that the parameter can possibly take. Each interval represents a set of possible values that a particular entity or variable may assume, without any a-priori assumption about exact value, probability measure or grade of membership. In particular, intervals of possibilities (the term was coined by Babad & Berliner (1994)) should be used whenever decision variables can assume different values, but a probability measure on these values is not available or justifiable. In reality, inexactness of this kind can be cited in countless numbers. Since the last decade, literatures on the *imprecise and uncertain data modeling* show that many a researchers in the fields of economics, engineering and natural scientific branches have taken interest in analyzing different real life situations where interval numbers appear as parameters (to quote a few, e.g., Inuiguchi & Kume (1991), Tong (1994), Das et al (1999), Tang et al (2000), Ida (2003), Giove et al (2006)). However, in the fields of decision problems its use has not much attended yet as it merits.

1.4. Interval Arithmetic: Notation and relevant preliminaries

All lower case letters denote real numbers and all upper case letters denote the interval numbers. An interval is defined as follows:

$$A = [a_L, a_R] = \{t \mid a_L \le t \le a_R, t \in \Re\},$$

where, a_L, a_R are respectively the left limit and right limit of the interval A on the real line $\Re$. If $a_L = a_R = a$, then $A = [a, a]$ is a real number.

Interval A, alternatively, is represented by its mid-point $m(A)$ and the half-width (or, simply be termed as 'width') $w(A)$ as

$$A = \langle m(A), w(B) \rangle,$$

where,

$$m(A) = \tfrac{1}{2}(a_L + a_R)$$

and

$$w(A) = \tfrac{1}{2}(a_R - a_L).$$

In literature, $m(A)$ has also been referred as *mean, center, central value* or *expected value* of interval A (Ishibuchi & Tanaka (1990)). Babad & Berliner (1994) defined $m(A)$ as the *plausible value*, around which an interval A of possibilities exists. Similarly, $w(A)$ alternatively has been termed as the *spread* or *range* or *level of uncertainty* or *the extent of uncertainty* or simply, *uncertainty* of interval A. The limits a_L and a_R have also been defined as the lower and upper bounds or as the *minimum* and *maximum* value or as the *infimum* and *supremum* (Babad & Berliner (1994), Giove 2002) of an interval A. Okada & Gen (1994) gave the name as *pessimistic* and *optimistic* value. Schjaer-Jacobsen (1996, 2002) used the name *worst-* and *best-case* (WBC) of economic consequences for an interval parameter.

Definition: Let $* = \{+, -, \cdot, /\}$ be a binary operation on the set of real numbers. Then, the extended operation '$\circledast$' between two interval numbers A and B is defined as

$$A \circledast B = \{a * b \mid a \in A, b \in B\}.$$

For division it is assumed that $0 \notin B$ (Alefeld & Herzberger (1983)). This definition of interval arithmetic operations is given on a set-theoretic basis and emphasizes the fact that sum, difference, product, or quotient of two intervals is just the set of sums, differences, products, or quotients of pair of real numbers taking one from each of the two intervals.

If λ is scalar, then

$$\lambda \cdot A = \lambda \cdot [a_L, a_R] = \begin{cases} [\lambda a_L, \lambda a_R], & \text{for } \lambda \geq 0 \\ [\lambda a_R, \lambda a_L], & \text{for } \lambda < 0 \end{cases}$$

or, in alternative notation,

$$\lambda A = \lambda.\langle m(A), w(B)\rangle = \langle \lambda m(A), |\lambda| w(B)\rangle.$$

The extended addition $\oplus$ and the extended subtraction $\ominus$ thus become:

$$A \oplus B = [a_L + b_L, a_R + b_R],$$

$$A \ominus B = [a_L - b_R, a_R - b_L].$$

Also, the extended multiplication $\odot$ and extended division $\oslash$ are defined as follows:

$$A \odot B = \left[\min\{a_L b_L, a_L b_R, a_R b_L, a_R b_R\}, \max\{a_L b_L, a_L b_R, a_R b_L, a_R b_R\}\right],$$

$$A \oslash B = [a_L, a_R] \cdot [1/b_R, 1/b_L], \text{ provided } 0 \notin [b_L, b_R].$$

It should be noted, however, that the straightforward application of interval arithmetic leads to too wide intervals. For example, the difference between two identical intervals A is not equal to $[0, 0]$,

$$\begin{aligned} A \ominus A &= [a_L, a_R] - [a_L, a_R] \\ &= \langle m(A), w(A) \rangle - \langle m(A), w(A) \rangle \\ &= \langle 0, 2w(A) \rangle \neq [0, 0]. \end{aligned}$$

Inuiguchi & Kume (1991) described this subtraction as *possible subtraction* and defined another form of subtraction as *necessary subtraction* '$)-($' as

$$A \;)-(\; B = [a_L - b_L, a_R - b_R], \text{ provided } w(A) \geq w(B).$$

Obviously, $A \;)-(\; B \subseteq A \ominus B$, and using this *necessary subtraction*, we have,

$$A \;)-(\; A = [a_L - a_L, a_R - a_R] = [0, 0].$$

In literature, however, Inuiguchi & Kume (1991)'s *necessary subtraction* could find very limited application. One reason for that might be the nonconformity with the Zadeh's extension principle, which is considered as the foundation of interval as well as fuzzy mathematics.

Use of interval parameter is a simple and intuitive way to introduce data uncertainty for large and complex decision problems. The way of argument as well as the advantages in handling data uncertainty in the form of interval numbers has attracted many researchers. Consequently, the problem of crisp or fuzzy interval (number) ordering has become a matter of perennial interest, also because of its direct relevance to the practical modeling and optimization in the real world processes under imprecise and uncertain environment. However, till date one main dilemma in using interval data for decision problems is perhaps the choice of an appropriate interval order relation. Theoretically, crisp intervals and fuzzy numbers can only be partially ordered and hence is argued that they cannot be compared (Sevastjanov (2004)). However, when interval or fuzzy numbers are used in practical applications or when a choice has to be made among alternatives, the comparison is indeed needed.

When two intervals A and B are

1) non-overlapping (disjoint) i.e., $a_R < b_L$, (Fig. 1.7a)
2) partially overlapping i.e., $a_L < b_L \leq a_R < b_R$, (Fig. 1.7b)
3) fully overlapping, such that B is nested in A, i.e., $B \subseteq A$ (Figs. 1.7c, 1.7d & 1.7e).

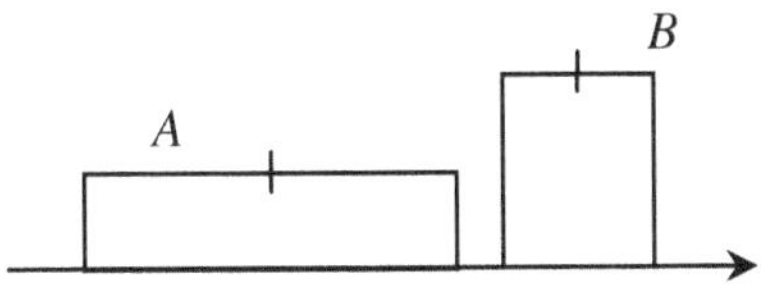

Fig. 1.7a. Non-overlapping intervals

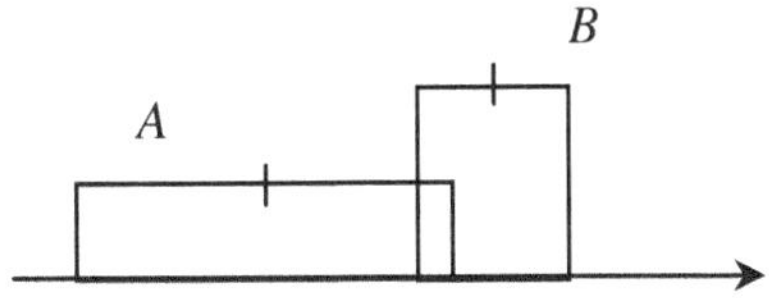

Fig. 1.7b. Partially overlapping intervals

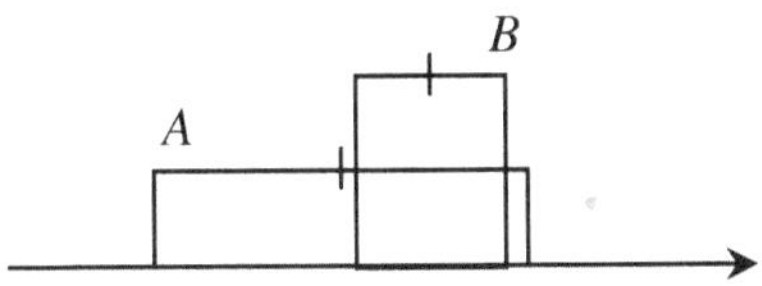

Fig. 1.7c. Overlapping intervals

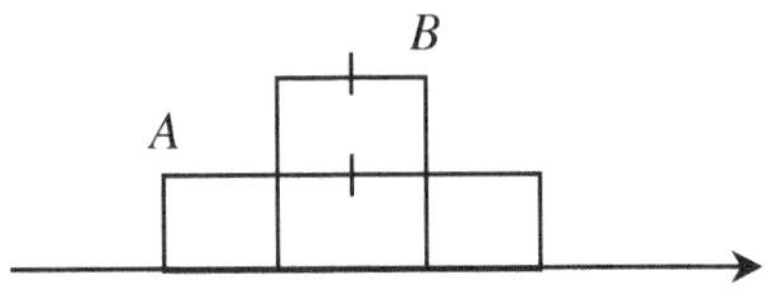

Fig. 1.7d. Overlapping intervals

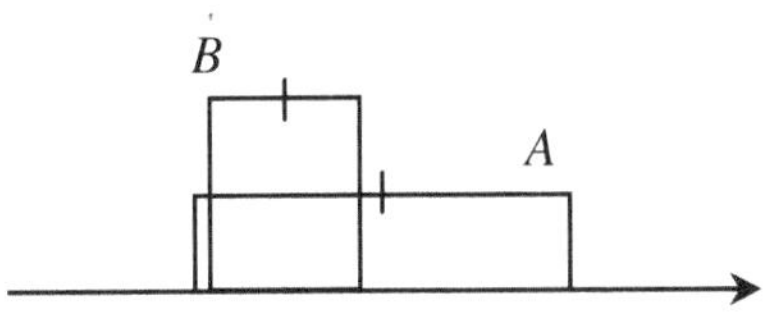

Fig. 1.7e. Overlapping intervals

Case 1 and case 2 show straightforward cases of ranking when A precedes B (Fig. 1.7a and 1.7b). However, the complexity arises in case 3. In literature, the debate in ranking is initiated from this point.

An *interval equality* is defined as

$$A = B, \text{ if } a_L = b_L \text{ and } a_R = b_R ,$$

(Moore (1966, 1979), Alefeld & Herzberger (1983), Alefeld & Mayer (1996)). However, there is difference between equality and indistinctiveness (or identity). The above relation should be read as *interval identity* or *interval indistinguishability* or *interval equivalence*, i.e., $A \equiv B$. The relation between A and B, such that

$$m(A) = m(B) \text{ and } w(A) \neq w(B),$$

is defined as equality in the discussion of spatial relationship between the two intervals (Kundu (1997)). In an earlier work, Ishibuchi & Tanaka (1990) defined a preference relation between such A and B, where one is nested just in the middle of the other interval. Therefore, when we try to define useful concepts like equality, similarity, indistinctiveness (identity or indistinguishability) or fuzzy equivalence, it is important first to clarify the semantic background and then introduce the definitions (Klawonn (2003)).

A major portion of this book is devoted to deal with the notion of equality and inequality between any pair of interval numbers and for finding a suitable and executable ranking strategy between them.

1.5. The motivation and organization of the book

The book consists of nine chapters including the present introductory chapter. The motivation and the content of each chapter are described briefly as follows.

A brief survey on the development of the interval comparing schemes starting from the work of Moore (1966) has been given in Chapter 2. The-state-of-the-art discussed in this book includes possibly all relevant documents in this subject. Comparison of different ranking methods for interval numbers is addressed to spot the lacunae. In our opinion, mainly Ishibuchi & Tanaka (1990)'s and Kundu (1997)'s remarkable works created space for further research. However, the study here discusses the ranking schemes published before 2000-01 and the rest of the part is discussed at the end of Chapter 4.

For comparing any two interval-numbers, Sengupta & Pal (2000, 2001) introduced two ranking methods of which the 1st one is called the *acceptability index* or, $\mathscr{A}$*-index*, in short. In Chapter 3, the definition and various properties of this index and some numerical illustrations have been given. An interval linear programming problem is defined here as an extension of the classical linear programming problem to an inexact environment. On the basis of $\mathscr{A}$-index, equality and inequality constraints involving interval coefficients are interpreted and reduced to their satisfactory crisp equivalent forms and a satisfactory solution of the optimization problem is defined. Next, efficiency of this methodology is discussed with a numerical example (Tong (1994)).

Chapter 4 presents a discussion on the strength and limitation of $\mathscr{A}$-index over the other approaches. $\mathscr{A}$-index is described here as a value judgment index for comparing any two real intervals. As a preference index, this index has also been called *an optimistic DM's preference index*. Next, the 2nd index – the *fuzzy preference ordering* between any two interval-attributes (say, profit, cost of production etc.) – has been introduced in this chapter in order to discuss a pessimistic or risk-averse DM's point of view. Different DMs' choices with different degrees of pessimism/optimism have also been considered and explained with suitable numerical example. Choice of an interval for a maximization and that for a minimization setup are dealt separately. Efficiency of these indices over the other ranking strategies has been summed up in a tabular set-up. Finally this chapter includes a comparative study on the other prominent ranking schemes that come after 2000-01.

This book next presents an algorithm of Shortest Path Problem with the connected arcs in a transportation network and its objective is to figure out the fastest route to travel from one node to other node. Each arc represents transportation time or cost in traveling and as a consequence, may fluctuate considerably depending on traffic condition, payload etc. In this problem, we define each arc as real intervals. Here, Dijkstra's algorithm has been redesigned where most of the parameters are inexact and given in terms of simple intervals. While solving the interval-valued shortest path problem, Okada & Gen (1994)'s algorithm generated some incomparable intervals as non-dominated solutions of the problem. However, we propose a methodology that considers *fuzzy preference ordering* of intervals (already defined in chapter 4) and it gives more clear and convincing decision than that of Okada & Gen (1994).

In Chapter 5, based on *fuzzy preference ordering*, a procedure of choosing a preferred minimum from a set of n intervals is defined first with suitable illustration of a numerical example. A shortest path problem with the connected arcs in interval data is defined and the extended Dijkstra's algorithm is presented and discussed next. In order to compare results obtained by Okada & Gen (1994), we also consider and discuss another large problem to show the efficiency and effectiveness of our approach.

The *Travelling Salesman Problem* (TSP) is one of the classical discrete (combinatorial) optimization problems, encountered in Operations Research (Lawler et al (1985), Clifford & Siu (1995)). Let us suppose that a salesman visits n cities (or nodes) cyclically. In one tour he visits each city just once, and finishes up where he started. In what order should he visit them to minimize the distance or cost travelled? The problem has some direct importance, since quite a lot of practical applications can be put in this form.

Chapter 6 describes such a TSP using interval arcs. The salesman here wants to visit a certain number of cities allotted to him, where the *cost* or *distance* or *time* of journey between every pair of cities are considered as interval numbers as these may fluctuate considerably depending on traffic condition, payload etc. The objective of this work is to select such a route that starts from his home city, passes through each city once and only once, and returns to his home city at lowest possible cost (or in least time or in the shortest possible distance). In this problem, we define each arc in terms of real intervals.

The problems of distribution of goods from manufacturer to customer are generally described under a common heading, *Transportation Problem* (TP). The TP, originally developed by (Hitchcock (1941)), can be used when a firm tries to decide where to locate a new facility. Good financial decisions concerning facility location also attempt to minimize total transportation and production costs for the entire system. Moreover, many problems are there not exactly being called the TP but can be modelled alike.

The aim of chapter 7 is to define a Multi-objective Interval-valued Transportation Problem and describe a procedure of getting a satisfactory compromise solution. In any inexact programming problem, before getting into the solution procedure, the first task comes to the front is to explain the meaning of the *objective* defined in the said inexact environment. Henceforth, maximization / minimization of an Interval Objective Function (IOF) is interpreted and finally a composite objective function is defined and the problem is solved.

Distinguishable features of this work are:

(1) that the methodology is completely based on the principles of interval arithmetic and on the analysis of interval order relation.

(2) that it can reflect decision maker's pessimistic or optimistic bias of different level in achieving the compromise solution.

(3) that for an ITPMPF (Interval-valued Transportation Problem with Multiple Penalty factors) with a number of IOFs, only a few steps are sufficient to yield the compromise solution of the said problem. For having more hold into the problem, one may try to have his desired satisfactory solution by changing objective weight or the optimism / pessimism parameter.

In these contexts the methodology given in chapter 7 creates greater impact in efficiency and effectiveness than that of the previous work (Das et al (1999)) from which the numerical example have been taken and studied.

Chapter 8 describes Fuzzy Preference based TOPSIS (Technique for Order Preference by Similarity to the Ideal Solution) for Interval Multi-criteria Decision Making. In almost all real-world decision making problems multiplicity of criteria for judging the alternatives is pervasive. That is, for many such problems a DM wants to solve a Multiple Criteria Decision-Making (MCDM) problem, which can concisely be expressed in matrix format as

$$D = \begin{array}{c} \\ A_1 \\ A_2 \\ \vdots \\ \vdots \\ A_m \end{array} \begin{array}{c} \begin{array}{cccccc} C_1 & C_2 & \cdots & \cdots & C_n \end{array} \\ \begin{bmatrix} x_{11} & x_{12} & \cdots & \cdots & x_{1n} \\ x_{21} & x_{22} & \cdots & \cdots & x_{2n} \\ \vdots & \vdots & \cdots & \cdots & \vdots \\ \vdots & \vdots & \vdots & \vdots & \vdots \\ x_{m1} & x_{m2} & \cdots & \cdots & x_{mn} \end{bmatrix} \end{array}$$

$$W = \begin{bmatrix} w_1 & w_2 & \cdots & w_n \end{bmatrix}$$

where, A_1, A_2, $\cdots$ A_m are possible alternatives among which DMs have to choose, C_1, C_2, $\cdots$ C_n are the criteria with which alternative performances are measured, x_{ij} is the rating of alternative A_i with respect to the criterion C_j and w_j is the weight of the criterion C_j.

In classical MCDM methods, the problem data given in terms of ratings and weights are all known preciously (Chen et al (1992), Dyer et al (1992)). The TOPSIS (Hwang & Yoon (1981)) is one of the most acceptable approaches in solving classical MCDM problems. TOPSIS defines an index called similarity or relative closeness to the *ideal solution* by combining the proximity to the *ideal solution* and the remoteness to the *negative-ideal solution.* Hence, TOPSIS selects an alternative with maximum relative similarity to the hypothetical ideal one.

An interval version of the traditional TOPSIS method is proposed in chapter 8. Giove (2002) described I-TOPSIS (Interval TOPSIS) approach using the *acceptability index* after suggesting some modification. But, being a value-based ranking index, the *acceptability index* has some limitation if it is used as a preference index. For attaining more flexibility and efficiency of DM's interval ranking strategy an extension of Giove (2002)'s work in the light of *Fuzzy Preference in interval order* has been proposed in chapter 8.

Chapter 9 gives the conclusion and future scope of research.

1.6. References

G. **Alefeld** & J. **Herzberger (1983)**, *Introduction to Interval Computations*, Academic Press, New York.

G. **Alefeld** & G. **Mayer (1996)**, Interval Analysis: theory and applications, *Journal of Applied Mathematics* 121 (1-2): 421-464.

A.D. **Athanassopoulos** & V.V. **Podinovski (1997)**, Dominance and potential optimality in multiple criteria decision analysis with imprecise information, *Journal of the Operational Research Society* 48 (2): 142-150.

Y.M. **Babad** & B. **Berliner (1994)**, *The use of intervals of possibilities to measure and evaluate financial risk and uncertainty*, AFIR Colloquium, Orlando, US. April 20-22, 1994 (www.actuaries.org/AFIR/collquia/Orlando/Babad_Berliner pdf).

C.A. **Bana e Costa** & P. **Vincke (1995)**, Measuring credibility of compensatory preference statements when trade-offs are interval determined, *Theory and Decision* 39 (2): 127 – 155.

A. **Bárdossy** & L. **Duckstein (1995)**, *Fuzzy rule-based modeling with applications to geophysical, biological and engineering systems*, CRC Press, Boca Raton, FL.

J.J. **Buckley (1987)**, The fuzzy mathematics of finance, *Fuzzy Sets and Systems* 21(3): 257-273.

Debjani **Chakraborty (1995)**, *Optimization in Imprecise and Uncertain Environment*, Ph. D. Thesis, Dept. of Mathematics, I.I.T. Kharagpur.

J.-L. **Chameau** & J.C. **Santamarina (1987)**, Membership functions I: comparing methods of measurement, *International Journal Approximate Reasoning* 1: 287-301.

A. **Charnes** & W.W. **Cooper (1959)**, Chance-constrained programming, *Management Science* 6: 73–79.

Shyi-Ming **Chen (2001)**, Fuzzy group decision making for evaluating the rate of aggregative risk in software development, *Fuzzy Sets and Systems* 118 (1): 75 – 88.

A.M.G. **Cornelissen**, J. van den Berg, W.J. Koops, M. Grossman & H.M.J. Udo (**2000**), *Assessment of sustainable development: a novel approach using fuzzy set theory*, ERIM Report Series Research in Management (ERS-2000-23-LIS), Rotterdam School of Management, The Netherlands.

S.K. **Das**, A. Goswami & S.S. Alam **(1999)**, Multiobjective transportation problem with interval cost, source and destination parameters, *European Journal of Operational Research* 117 (1): 100 – 112.

M. **Delgado**, J.L. Verdegay, M.A. Vila **(1989)**, A general model for fuzzy linear programming, *Fuzzy Sets and Systems* 29 (1): 21–29.

J.L. **Deng (1982)**, Control problems of grey systems, *Systems and Control Letters* 1 (5): 211–215.

D. **Dubois** & H. **Prade (1980)**, *Fuzzy sets and systems - Theory and applications*, Academic press, New York.

S. **Giove (2002)**, Interval TOPSIS for Multicriteria Decision Making, in: M. Marinaro & R. Tagliaferri (Eds.): WIRN VIETRI 2002, *Lecture Notes in Computer Science* 2486, Springer-Verlag Berlin Heidelberg: 56-63.

S. **Giove**, S. Funari & C. Nardelli **(2006)**, An interval portfolio selection problem based on regret function, *European Journal of Operational Research* 170 (1): 253-264.

M.M. **Gupta (1988)**, Cognition, Perception and Uncertainty, in: M.M. Gupta & T. Yamakawa (Eds.): *Fuzzy Logic in Knowledge-based Systems, Decision and Control*, Elsevier Science Publisher B.V., North Holland.

G.B. **Hazen (1986)**, Partial Information, Dominance, and Potential Optimality in Multiattribute Utility Theory, *Operations Research* 34 (2): 296-310.

G.H. **Huang**, B.W. Baetz & G.G. Patry **(1995)**, A grey integer programming approach for waste management planning, *European Journal of Operational Research* 83: 594–620.

M. **Ida (2003)**, Portfolio Selection Problem with Interval Coefficients, *Applied Mathematics Letters* 16 (5): 709-713.

M. **Inuiguchi** & Y. **Kume (1991)**, Goal Programming problem with Interval Coefficients and Target Intervals, *European Journal of Operational Research* 52 (3): 345 – 360.

P. **Kall (1982)**, Stochastic programming, *European Journal of Operational Research* 10: 125–130.

S. **Karmakar** & P.P. **Mujumdar** **(2006)**, An inexact optimization approach for river water -quality management, *Journal of Environmental Management* 81: 233–248.
P. **Kall** **(1982)**, Stochastic programming, *European Journal of Operational Research* 10: 125–130.

S. **Karmakar** & P.P. **Mujumdar** **(2006)**, An inexact optimization approach for river water -quality management, *Journal of Environmental Management* 81: 233–248.

A. **Kaufmann** & M.M. **Gupta** **(1985)**, *Introduction to Fuzzy Arithmetic—Theory and Applications*, Van Nostrand, Reinhold.

R.B. **Kearfott** & V. **Kreinovich** **(1996)**, *Applications of Interval Computations*, Kluwer Academic Publishers, Dordrecht, Netherlands.

F. **Klawonn** **(2003)**, Should fuzzy equality and similarity satisfy transitivity? Comments on the paper by M. De Cock & E. Kerre, *Fuzzy Sets and Systems* 133 (2): 175–180.

G.J. **Klir** & B. **Yuan** **(1995)**, *Fuzzy Sets and Fuzzy Logic: Theory and Application*, Prentice-Hall, New Jersey.

G.J. **Klir** & T.A. **Folger** **(1988)**, *Fuzzy Sets, Uncertainty, and Information*, Prentice-Hall, New Jersey.

B. **Kosko** **(1990)**, Fuzziness vs. probability, *International Journal of General Systems* 17 (1): 211–240.

S. **Liu** & Y. **Lin** **(1998)**, *An Introduction to Grey Systems: Foundation, Methodology and Applications*, IIGSS Academic Publishers, Pennsylvania.

W.A. **Lodwick** & K.D. **Jamison** **(2003)**, Special issue: interfaces between fuzzy set theory and interval analysis, *Fuzzy Sets and Systems* 135 (1): 1–3.

M.K. **Luhandjula** **(1983)**, Linear programming under randomness and fuzziness, *Fuzzy Sets and Systems* 10: 45 – 55.

M.K. **Luhandjula** **(1989)**, Fuzzy optimization: An appraisal, *Fuzzy Sets and Systems* 30 (3): 257–282.

D. **McNeill** & P. **Freiberger** **(1993)**, *Fuzzy Logic*, Simon and Schuster, New York.

K. **Miettinen** & P. **Salminen** **(1999)**, Decision-Aid for Discrete Multiple Criteria Decision Making Problems with Imprecise Data, *European Journal of Operational Research* 119 (1): 50-60.

R.E. **Moore** **(1979)**, *Method and Application of Interval Analysis*, SIAM, Philadelphia.

M. **Mukaidono** **(2001)**, *Fuzzy Logic for beginners*, World Scientific, New Jersey.

G. **Munda**, P. Nijkamp & P. Rietveld **(1992)**, *Multicriteria Evaluation and Fuzzy Set Theory: Applications in Planning for Sustainability*, Serie Research Memoranda - 68, Free University, Amsterdam.

H.T. **Nguyen** & V. **Kreinovich** **(2005)**, Optimization and Decision Making under Interval and Fuzzy Uncertainty: Towards New Mathematical Foundations, in: Cengiz Kahraman (ed.), *Fuzzy Set Techniques in Industrial Engineering*, Springer Verlag.

A.M. **Norwich** & I.B. **Türksen** **(1984),** A model for the measurement of membership and the consequences of its empirical implementation, *Fuzzy Sets and Systems* 12: 1-25.

S. **Okada** & M. **Gen (1994)**, Order Relation between Intervals and Its Application to Shortest Path Problem, *Japanese journal of Fuzzy Theory and Systems* 6 (6): 703 – 717. (In English).

Karl **Popper (1974)**, Autobiography of Karl Popper, in: P.A. Schilpp (Ed.): *The Philosophy of Karl Popper*, Open Court La Salle, Illinois: 1 – 181.

P. **Prodanovic** & S.P. **Simonovic (2002)**, Comparison of fuzzy set ranking methods for implementation in water resources decision making, *Canadian Journal of Civil Engineering* 29 (5): 692 – 701.

R. **Rommelfanger (1989)**, Interactive decision making in fuzzy linear optimization problems, *European Journal of Operational Research* 41 (2): 210–217.

T. **Santoso**, S. Ahmed, M. Goetschalckx & A. Shapiro **(2005)**, A stochastic programming approach for supply chain network design under uncertainty, *European Journal of Operational Research* 167 (1): 96 – 115.

J.C. **Santamarina** & J.-L. **Chameau (1987)** Membership functions II: trends in fuzziness and implications, *International Journal Approximate Reasoning* 1: 303-317.

H. **Schjaer-Jacobsen (1996)**, A new method for evaluating worst- and best-case (WBC) economic consequence of technological development, *International journal of Production Economics* 46-47: 241 – 250.

H. **Schjaer-Jacobsen (2002)**, Representation and calculation of economic uncertainties: Intervals, fuzzy numbers and probilities, *International journal of Production Economics* 78: 91 – 98.

J.K. **Sengupta (1981)**, *Optimal Decision under Uncertainty*, Springer, New York.

P. **Sevastjanov (2004)**, Interval comparison based on Dempster-Shafer theory of evidence, LNCS 3019 (2004) Springer-Verlag Heidelberg: 668 – 675.

R. **Slowinski (1986)**, A multicriteria fuzzy linear programming method for water supply systems development planning, *Fuzzy Sets and Systems* 19: 217–237.

T. **Taguchi**, T. Yokota & Mitsuo Gen **(1998)**, Reliability Optimal Design Problem with Interval Coefficients Using Hybrid Genetic Algorithms, *Computers and Industrial Engineering* 35 (1): 373-376.

S. **Talluri**, R. Narasimhana & A. Nair **(2006)**, Vendor performance with supply risk: A chance-constrained DEA approach, *Int. J. Production Economics.* 100 (2): 212 – 222.

J. **Tang**, D. Wang & R.Y.K. Fung **(2000)**, Fuzzy formulation for multi-product aggregate production planning, *Production planning & control* 11(7): 670 – 676.

S. **Tong (1994)**, Interval number and fuzzy number linear programming, *Fuzzy Sets and Systems* 66: 301–306.

I.B. **Türksen (1991)**, Measurement of membership functions and their acquisition, *Fuzzy Sets and Systems* 40: 5-38.

I.B. **Türksen (1999)**, Type I and Type II fuzzy system modeling, *Fuzzy Sets and Systems* 106: 11-34.

S. **Vajda (1972)**, *Probabilistic Programming*, Academic Press, New York.

J. **Vehí**, J. Armengol & L.E. Müjica (**2002**), A Review on Interval Methods for Adaptive Systems, *Second EUNITE Annual Symposium*, Albufeira, Algarve, Sep 18, 2002.

M. **Weber (1987)**, Decision making with incomplete information, *European Journal of Operational Research* 28 (1): 44 – 57.

Y. **Yang** & R.I. **John (2003)**, Grey systems and interval valued fuzzy sets, *Proceedings of European Society for Fuzzy Logic and Technology - 2003*, Zittau, Germany, 10–12 September 2003, pp. 193–197.

J. **Yen** & R. **Langari (1999)**, *Fuzzy Logic – Intelligence, Control and Information*, Pearson Education, Inc.

L.A. **Zadeh (1962)**, From Circuit Theory to System Theory, *Proceedings of the Institute of Radio Engineers* 50: 856-865.

L.A. **Zadeh (1965)**, Fuzzy Sets, *Information and Control* 8(3): 338–352.

L.A. **Zadeh (1978)**, PRUF – A meaning representation language for natural languages, *Int. Journal of Man-Machine Studies* 10: 395 – 460.

H.J. **Zimmermann (1991)**, *Fuzzy Set Theory and its Application,* Kluwer Academic Publishers, Boston/Dordrecht/London.

Chapter 2

On Comparing Interval Numbers: A Study on Existing Ideas

2.1. Introduction

The problem of comparing imprecise or uncertain quantities has been attempted by many researchers. In literature, as much as 40 schemes for ranking fuzzy quantities have been presented during last three decades with a lot of debates and contradictory claims. Inconsistency of the ranking orders among different approaches and the debates thereon were due to the different criteria of selections. Given two imprecise quantities, it is assumed that a set of precisely defined decision objectives are needed in order to decide, which of the two precedes or ranks above the other, or, which of the notions – the notion of *preference* or the notion of *greater than* (or *less than*) – is to be given the priority, or, whether an optimally selected (greatest/lowest) set satisfies an intuitive notion of being the most preferred (Wang & Kerre (1996, 2001a, 2001b)). In fuzzy literatures, we find some remarkable research papers that categorize and compare ranking strategies implicitly or explicitly on the basis of some criteria, such as the distinguishability (Bortolan & Degani (1985)), the rationality (Nakamura (1986)), and the fuzzy or linguistic presentation (Tong & Bonissone (1980) and Delgado et al. (1988)).

A. Sengupta and T.K. Pal: Fuzzy Preference Ordering, STUDFUZZ 238, pp. 25–37.
springerlink.com

In interval ranking there are not much contradictions and debates as we witnessed in fuzzy ranking. A possible reason might be the timing of the concentration of the relevant publications on interval ranking. In our opinion the most outstanding input in this field came out of Ishibuchi & Tanaka (1990)'s work. After the debate on fuzzy ranking had been almost over, this interval ranking methodology appeared as a fair blend of optimality and rational choice of the DM. What it can do has not been challenged by most of the researchers. What it can not do has only been investigated thereafter keeping the basic findings of Ishibuchi & Tanaka (1990) intact as the fundamental. Research on interval ranking thereafter has been aimed towards finding a complete and robust methodology that can be applied to any two real interval attributes.

Here we are going to study the development of the interval comparing schemes starting from the work of Moore (1966). Intension of this part of our study is to include possibly all of the relevant works published till the recent past and establish a logical path of development with comparative analyses among them. Such a discussion on some diversified ideas and their comparative study on a single platform is not at all an easy task. Unintended error of non-inclusion of any important research work is quite possible and is highly regretted.

2.2. Criteria for comparing interval numbers

In literature there exist two different types of notions to represent an interval. The first one assumes that an interval number is a non-fuzzy subset of the real line with a precisely defined domain on it. An interval $A = [a_L, a_R]$ is defined as

$$A = [a_L, a_R] = \{t \mid a_L \leq t \leq a_R, \; a_L, a_R \in \Re\}$$

It is an ordered pair of real numbers. This representation can also assume the following form where $m(A)$ and $w(A)$ are the center and radius of a disk mapped on the real line $\Re$,

$$A = \langle m(A), w(A) \rangle = \{t' \mid m(A) - w(A) \leq t' \leq m(A) + w(A), \; m(A), w(A) \in \Re\}$$

where, A can be defined again as

$$A = [m(A) - w(A), \; m(A) + w(A)].$$

This interval can be represented in another way by its *characteristic function*, $\forall u \in A$ on $\Re$, as

$$\mu_A(u) = \begin{cases} 1, & \text{if } a_L \leq u \leq a_R \\ 0, & \text{otherwise} \end{cases}$$

Here t and u are distributed within a_L and a_R, but the distribution function is not known. Obviously, the notion of interval numbers provides a tool for representing a real number by specifying its lower and upper endpoints. This representation scheme is very useful in situations where we are unable to obtain a precise measurement of a physical quantity (i.e., inexact experimental measurement), or where we cannot store a real number with sufficient precision (i.e., insufficient representation), or where a human expert's subjective opinion is needed for an objective transcription (Yao (1993)).

A *grey number* or *interval grey number* is defined in the literature as a similar sort of number whose exact value is unknown but a range within which the value lies is known (Liu & Lin (1998), Yang & John (2003)). Unknown distribution within a known pair of bounds is the only available information about the number (Huang et al (1993)). Though grey systems theory has its own place in the literature that deals in the uncertainty issue, the application papers using grey number use conventional interval ranking schemes of the above set-theoretic form (Karmakar & Mujumdar (2006a, 2006b), Zhou et al (2008)).

Whereas the other representation assumes the notion that there exist two closed intervals A and B on the real line, such that random variables $x \in A$ and $y \in B$ are uniformly and independently distributed on their respective intervals. For example, $P(x \leq y)$ represents conventional probability of $x \leq y$ and may be interpreted as probability of $A \leq B$ (Nakahara et al (1992), Kundu (1997)).

The former definition assumes a set-theoretic representation whereas the latter assumes a probabilistic representation. In interval literature, the former definition has been used the most.

An interval number comparing scheme can be viewed either as value based or preference based. A value-based ordering scheme (which emphasizes only on the notion of *greater/less than*) assumes a consistent and coherent total order in the set of intervals and should have a cardinal representation. Here consistency may be interpreted as asymmetry of the form: if an interval A is ranked greater than interval B, then B can not be greater than A. Similarly, the term coherent is interpreted as the transitivity property: if A, B and C are three intervals and if A is greater than B and B is greater than C then A is greater than C.

On the other hand, a preference-based ordering scheme assumes a sort of partial order where strict and fuzzy orders co-exist. This ranking scheme should be able to represent rational human behaviour or intuition in terms of consistency and coherence in preference ordering. The ranks obtained from the value and preference based schemes may contradict. The goodness of any comparing scheme, however, lays in its inherent blend of numerical optimality and preference (rational choice which is intrinsically fuzzy in nature) criteria.

2.3. Interval comparing schemes

2.3.1. *Set theoretic approaches*

An extensive research and wide coverage on interval arithmetic and its applications can be found in Moore (1966, 1979)'s works. Here we find two transitive order relations defined over intervals $A = [a_L, a_R]$ and $B = [b_L, b_R]$: the first one as an extension of '<' on the real line as

$$A < B, \text{ iff } a_R < b_L$$

and the other as an extension of the concept of set inclusion i.e.

$$A \subseteq B \text{ iff } a_L \geq b_L \text{ and } a_R \leq b_R.$$

These order relations cannot explain ranking between two overlapping intervals. The first order relation defines the simplest case of two non-overlapping intervals. Whereas the second one, appears here as the extension of the set inclusion, describes only the condition that the interval A is nested in B; but it cannot order A and B neither in terms of value nor in terms of preference. Hence, there was a need to develop a definition for comparing any two interval numbers even when one is partially/fully nested in another.

Ishibuchi & Tanaka (1990) approached the problem of ranking two interval numbers more prominently. In this approach, in a maximization problem if intervals A and B are two, say, profit intervals, then maximum of A and B can be defined by an order relation $\leq_{LR}$ between A and B as follows:

$$A \leq_{LR} B \text{ iff } a_L \leq b_L \text{ and } a_R \leq b_R,$$

$$A <_{LR} B \text{ iff } A \leq_{LR} B \text{ and } A \neq B.$$

The order relation $\leq_{LR}$ defines that if both the lower and upper limit of an interval is higher than that of another interval, then the former is higher valued interval than the latter. So, for a maximization problem, say, if A and B be two interval-valued profits, then interval B is higher valued and hence preferred to A. Similarly, for a minimization problem, if A and B be two interval costs, then interval A is lower valued and hence preferred to B.

When one interval is nested in another and the aforementioned $\leq_{LR}$ is not applicable, Ishibuchi & Tanaka (1990) suggested another order relation $\leq_{mw}$ as follows:

$$A \leq_{\mathrm{mw}} B \quad \text{iff} \quad m(A) \leq m(B) \ \text{ and } \ w(A) \geq w(B),$$

$$A <_{\mathrm{mw}} B \ \text{ iff } \ A \leq_{\mathrm{mw}} B \ \text{ and } \ A \neq B.$$

The second order relation $\leq_{\mathrm{mw}}$ defines that if the central value of an interval is higher than that of another interval as well as if the latter is a wider interval, then the former interval is said to be preferred if the problem were to choose maximum from among the two. However, for a minimization problem, say, for A and B being the interval costs, if A is to be chosen as a preferred minimum, then the condition is given as follows:

$$A \leq^{*}_{\mathrm{mw}} B \quad \text{iff} \quad m(A) \leq m(B) \ \text{ and } \ w(A) \leq w(B),$$

$$A <^{*}_{\mathrm{mw}} B \ \text{ iff } \ A \leq^{*}_{\mathrm{mw}} B \ \text{ and } \ A \neq B.$$

These $\leq_{\mathrm{mw}}$ and $\leq^{*}_{\mathrm{mw}}$ relations reflect human intuition or rationality

Ultimately the order relations $\leq_{\mathrm{LR}}$ and $\leq_{\mathrm{mw}}$, for a maximization problem, have been summed up in a singular form as follows:

$$A \leq_{\mathrm{Lm}} B \quad \Leftrightarrow \quad a_{\mathrm{L}} \leq b_{\mathrm{L}} \ \text{and} \ m(A) \leq m(B),$$

and for a minimization problem as

$$A \leq_{\mathrm{mR}} B \quad \Leftrightarrow \quad a_{\mathrm{R}} \leq b_{\mathrm{R}} \ \text{and} \ m(A) \leq m(B).$$

Both of the above order relations $\leq_{\mathrm{LR}}$ and $\leq_{\mathrm{mw}}$ are anti-symmetric, reflexive and transitive and hence, define partial ordering between intervals. But they did not compare the pairs of intervals for which both $\leq_{\mathrm{LR}}$ and $\leq_{\mathrm{mw}}$ fail. As we conceive, here lies a drawback in their approach. They concentrated more on preference ordering, particularly on strict preference ordering, but less on the ranking in terms of value. While considering preference ordering between two inexact attributes represented by two intervals, consideration of a weak preference ordering between two intervals is more significant from a decision making point of view, particularly when a decision is to be made in an inexact environment. From this point of view their approach is incomplete and hence looses its significance.

The ranking approach of Ishibuchi & Tanaka (1990) is limited from another point of view. Let us take an example: there exists a constraint $Ax \leq B$ in a decision model (as has been stated in the optimization problem given in the next chapter), where, both of the coefficients, A and B are any two intervals. In this context, to exploit the decision variable x exhaustively within the scope of the interval-valued constraint, an exclusive ranking strategy, i.e., ranking left-hand and right-hand sides of the inequality in terms of value is required. Here a value based ranking scheme is indeed needed. Can any one or both of $\leq_{LR}$ and $\leq_{mw}$ be utilized for this purpose? Perhaps, not.

Okada & Gen (1994) extended Ishibuchi & Tanaka (1990)'s methodology and defined an interface between partial and total orderings. Their approach (Okada & Gen (1994)) tried to reduce the number of incomparable pairs by trial-and-error from among a set of intervals. When a greater interval is to be preferred, an order relation $\leq_{\alpha,\beta}$ for α and β (such that $0 \leq \alpha \leq \beta \leq 1$) is defined as follows:

$$A \leq_{\alpha,\beta} B \iff \begin{cases} m(B) - m(A) \geq \alpha\,(w(B) - w(A)) \\ m(B) - m(A) \geq \beta\,(w(B) - w(A)) \end{cases}$$

Now, the both of the relations,

$$A \leq_{\alpha,\beta} B \iff \begin{cases} m(B) - m(A) \geq 0, \text{ if } \alpha = \beta = 0, \\ a_L \leq b_L, \text{ if } \alpha = \beta = 1. \end{cases}$$

define total order, whereas $A \leq_{0,1} B \iff A \leq_{Lm} B$ leads to the findings same as Ishibuchi & Tanaka (1990) and defines a partial order relation.

While dealing with an interval-valued transportation problem, Chanas & Kuchta (1996a, 1996b) did not consider any separate ranking index but restructured Ishibuchi & Tanaka (1990)'s different approaches into a single scheme using a sort of bi-parametric cut of the comparing intervals. We have some analytical discussion on this ranking approach from an application point of view in Section 7.3.2.

2.3.2. *Probabilistic approaches*

Using probabilistic concept, Nakahara et al (1992) defines probability of the interval inequality $P(A \leq B)$ for any two intervals $A = [a_L, a_R]$ and $B = [b_L, b_R]$, as follows:

$$P(A \leq B) = P(a \leq b),$$

where, a and b are the random variables uniformly distributed on A and B respectively and $P(a \leq b)$ is the conventional probability of $a \leq b$.

Nakahara et al (1992) studied probability of $A \leq B$ for the following six positions of A and B on the real line (see Fig. 2.1),

(i) $a_L \leq a_R \leq b_L \leq b_R$

(ii) $a_L \leq b_L \leq a_R \leq b_R$

(iii) $a_L \leq b_L \leq b_R \leq a_R$

(iv) $b_L \leq a_L \leq a_R \leq b_R$

(v) $b_L \leq a_L \leq b_R \leq a_R$

(vi) $b_L \leq b_R \leq a_L \leq a_R$

and defined the probability value for all the six cases in a *generalized format* as

$$P(A \leq B) = \frac{w - a_L}{a_R - a_L} + \frac{v - w}{(a_R - a_L)(b_R - b_L)}(b_R - \frac{v + w}{2}),$$

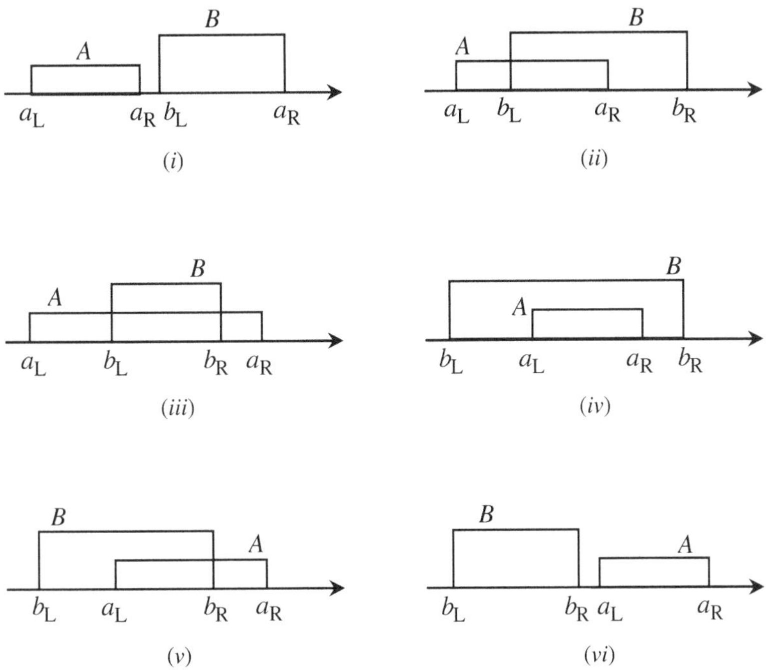

Fig. 2.1. Six combinations of intervals *A* and *B* on the real line

where,

$$w = \min\left\{\max\left\{a_L, b_L\right\}, a_R\right\},$$

$$v = \max\left\{\min\left\{a_R, b_R\right\}, a_L\right\}.$$

For a certain position of A and B on the real line, say for position (ii) as is given above, we have,

$$P(A \le B) = \frac{b_L - a_L}{a_R - a_L} + \frac{a_R - b_L}{(a_R - a_L)(b_R - b_L)}(b_R - \frac{a_R + b_L}{2}).$$

However, the way the generalized format of $P(A \le B)$ was conceived is not very simple to visualize. To be more specific, the interpretation or visualization of w and v is not easy. Instead, it is easier to define the case-by-case individual graphical construction of $P(A \le B)$. Here we try to frame the examples graphically for each of the six combinations of intervals A and B as stated above (see Fig. 2.1 and Fig. 2.2). Each of the six cases given in Fig.2.1 is explained in Fig. 2.2, where $P(A \le B)$ is interpreted as the proportion of the shaded area in the area delimited by the bold-lined rectangular block. Hence, apart from Nakahara et al (1992)'s construction, directly we can compute the value of $P(A \le B)$. For example, for position (ii), we get,

$$P(A \le B) = \frac{(b_L - a_L)(b_R - b_L) + (a_R - b_L)(b_R - a_R) + \frac{1}{2}(a_R - b_L)^2}{(a_R - a_L)(b_R - b_L)},$$

or, in another format,

$$P(A \le B) = 1 - \frac{\frac{1}{2}(a_R - b_L)^2}{(a_R - a_L)(b_R - b_L)}.$$

Nakahara et al (1992) mainly tried to investigate linear programming problem with interval coefficients. But, in our opinion, their most significant contribution in specific was the proposition of a probabilistic concept of constraints involving interval data. Earlier in Charnes & Cooper (1959)'s chance-constraint (stochastic) programming, we found a similar sort of approach.

Interval numbers were viewed as *independent uniform distributions* by other authors too. While considering the problem of defining fuzzy spatial relationships with the notions *left, right, above* etc. between two objects in a 2-dimensional scene in image processing, Kundu (1997) studied various aspects of the fuzzy con-

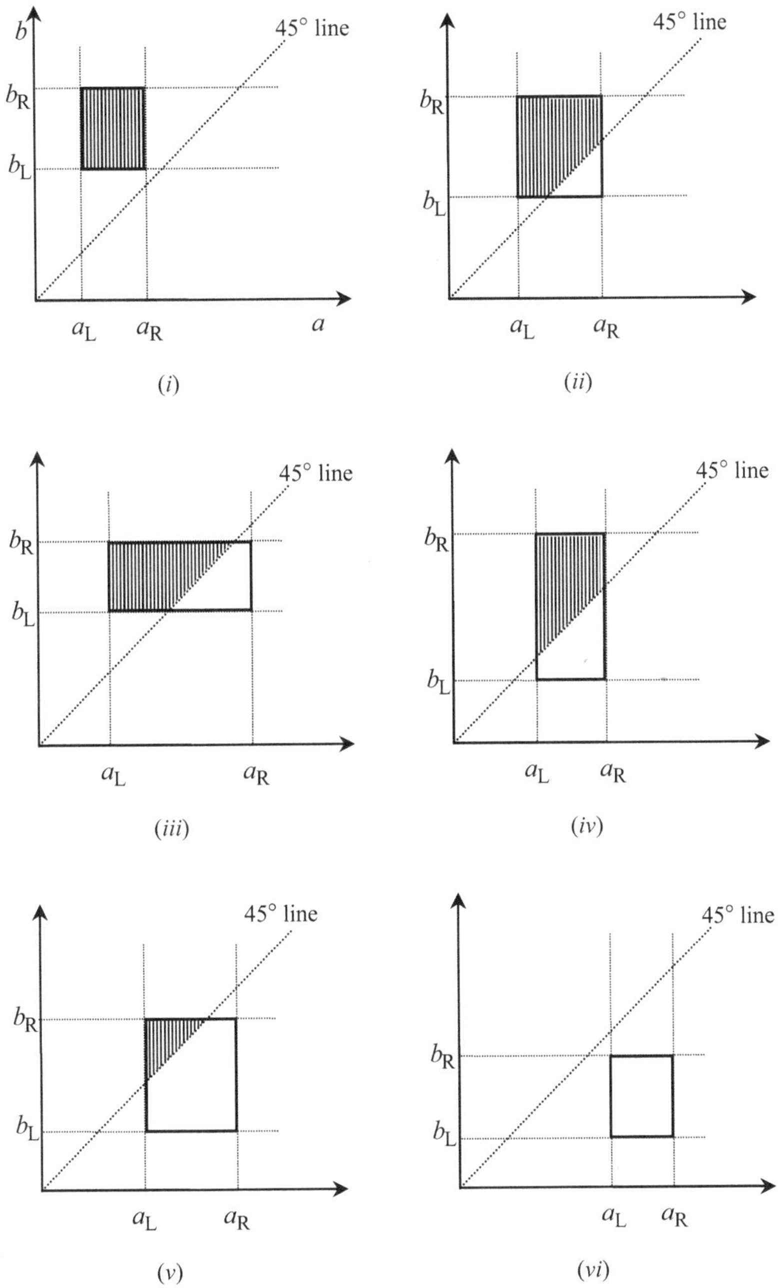

Fig. 2.2. Graphical representation of the cases given in Fig. 2.1.

cept 'left of an interval'. Using probability-based construction, Kundu (1997) defined a *fuzzy preference relationship* between any two intervals A and B on the real line by the formula

$$\text{Left}(A,B) = \max\{0, P_{AB}(x < y) - P_{AB}(x > y)\},$$

where, $P_{AB}(x < y)$ denotes the probability that $x < y$, given that $x \in A$ and $y \in B$ are uniformly and independently distributed in the intervals A and B. Similarly, Right(A, B) can be developed by using $P_{AB}(x > y) - P_{AB}(x < y)$. *Leftness* and *rightness* are interpreted as follows:

If $\text{Left}(A,B) > 0$, then in terms of value, A is left to B.

If $\text{Right}(A,B) > 0$, then in terms of value, A is right to B.

Visually the construction of *Leftness* or *rightness* relationship bears some resemblance to the findings of Nakahara et al (1992). But from the perspective of simplicity of concept and analysis, in our opinion, Kundu's work is more significant in the application and theoretical studies of ordering interval numbers. Kundu's *fuzzy leftness* (or *rightness*) *relationship* directly defines a preference ranking scheme by constructing a fuzzy membership function of an interval spatially related (by left or by right) to another interval on the real line. The author (Kundu (1997)) claimed that the selection of the least (or the most) preferred item from a set of alternatives can be made by using Left (A, B) (or Right (A, B)).

Let us take a simple example of choosing the best minimum from among the cost intervals

$$A = [20, 40] = \langle 30, 10 \rangle,$$

$$B = [22, 32] = \langle 27, 5 \rangle$$

$$C = [26, 34] = \langle 30, 4 \rangle$$

Certainly the best minimum among the alternatives does not mean the interval lowest in value. We have the solutions as follows:

$$\text{Left}(A,B) = 0, \qquad \text{Left}(B,A) = 0.3,$$

$$\text{Left}(B,C) = 0.55, \qquad \text{Left}(C,B) = 0,$$

$$\text{Left}(A,C) = 0, \qquad \text{Left}(C,A) = 0.$$

Therefore, B is the best choice, and next in the absence of B, both A and C are optimal choices.

Although Kundu's approach is intuitively appealing and can solve the Ishibuchi & Tanaka (1990)'s incomparable situation, the main disadvantage in this approach is that it is not consistent with a rational DM's preference pattern (consider the second part of the above result). Both A and C are having the same mean but A is more uncertain than C is. To a rational decision maker (even to a risk lover there is no other point of *attraction* in A except higher uncertainty), hence, C should be preferred to A.

In order to deal with the incomparable cases of nested intervals (Ishibuchi & Tanaka (1990)) and the intuitive inconsistency arising out the cases of ranking equi-centered intervals (Kundu (1997)), two ranking schemes, viz., *acceptability index* and *fuzzy preference ordering* for any two interval-numbers, were introduced by Sengupta & Pal (2000, 2001). In the next chapter, various properties of acceptability index (or $\mathscr{A}$-index) and its application in Interval Linear Programming Problem have been given in detail. In Chapter 4, after discussing advantage and disadvantage of $\mathscr{A}$-index, *fuzzy preference ordering* is introduced for maximization and minimization set-up. A detail comparative study on all the relevant ranking schemes has been given there.

There are also a few ranking schemes appeared and discussed from some application point of view after 2000-01. Giove (2002) used the acceptability index in an application after suggesting some modification. We have given our study with a numerical analysis on this issue in the Sections 8.2 and 8.3. Among the other ranking schemes, the most prominent schemes (Wang et al (2005), Sun & Yao (2008), Jiang et al (2008)) are included in this study in Section 4.6.1 in order to associate them in a time-frame as we conceive.

2.4. References

G. **Bortolan** & R. **Degani** **(1985)**, A review of some methods for ranking fuzzy subsets, *Fuzzy Sets and Systems* 15: 1-19.

S. **Chanas** & D. **Kuchta** **(1996a)**, A concept of solution of the Transportation Problem with fuzzy cost coefficient, *Fuzzy Sets and Systems* 82 (3): 299 – 305.

S. **Chanas** & D. **Kuchta** **(1996b)**, Multiobjective programming in optimization of interval objective functions – A generalized approach, *European Journal of Operational Research* 94 (3): 594 – 598.

A. **Charnes** & W.W. **Cooper** **(1959)**, Chance-constrained programming, *Management Science* 6: 73–79.

M. **Delgado**, J.L. Verdegay and M.A. Vila **(1988)**, A procedure for ranking fuzzy numbers using fuzzy relations, *Fuzzy Sets and Systems* 26: 49-62.

S. **Giove** **(2002)**, Interval TOPSIS for Multicriteria Decision Making, in: M. Marinaro & R. Tagliaferri (Eds.): WIRN VIETRI 2002, *Lecture Notes in Computer Science* 2486, Springer-Verlag Berlin Heidelberg: 56-63.

H. **Ishibuchi** & H. **Tanaka (1990)**, Multiobjective programming in optimization of the interval objective function, *European Journal of Operational Research* 48 (2): 219 –225.

C. **Jiang** , X. Han, G.R. Liu & G.P. Liu **(2008)**, A nonlinear interval number programming method for uncertain optimization problems, *European Journal of Operational Research* 188: 1–13.

S. **Karmakar** & P.P. **Mujumdar (2006a)**, An inexact optimization approach for river water-quality management, *Journal of Environmental Management* 81: 233–248.

S. **Karmakar** & P.P. **Mujumdar (2006b)**, Grey fuzzy optimization model for water-quality management of a river system, *Advances in Water Resources* 29: 1088–1105.

S. **Kundu (1997)**, Min-transitivity of fuzzy leftness relationship and its application to decision making, *Fuzzy Sets and Systems* 86 (3): 357–367.

S. **Liu** & Y. **Lin (1998)**, *An Introduction to Grey Systems: Foundation, Methodology and Applications.* IIGSS Academic Publishers, Pennsylvania.

R.E. **Moore (1966)**, *Interval Analysis*, Prentice-Hall, Englewood Cliffs, New Jersey.

R.E. **Moore (1979)**, *Method and Application of Interval Analysis*, SIAM, Philadelphia.

Y. **Nakahara,** M. Sasaki & M. Gen **(1992)**, On the linear programming problems with interval coefficients, *Computers and Industrial Engineering* 23(1-4): 301-304.

K. **Nakamura (1986),** Preference relations on a set of fuzzy utilities as a basis for decision making, *Fuzzy Sets and Systems* 20: 147-162.

S. **Okada** & M. **Gen (1994)**, Order Relation between Intervals and Its Application to Shortest Path Problem, *Japanese journal of Fuzzy Theory and Systems* 6 (6): 703 – 717. (In English).

A. **Sengupta** & T.K. **Pal (2000)**, On comparing interval numbers, *European Journal of Operational Research* 127: 28–43.

A. **Sengupta**, T.K. Pal & D. Chakraborty (**2001**), Interpretation of inequality constraints involving interval coefficients and a solution to interval linear programming. *Fuzzy Sets and Systems* 119: 129–138.

H. L. **Sun** & W.X. **Yao (2008)**, The basic properties of some typical systems' reliability in interval form, *Structural Safety* 30: 364–373.

R.M. **Tong** & P.P. **Bonissone (1980)**, A linguistic approach to decision making with fuzzy sets, *IEEE Trans. Systems Man Cybernet.* 10: 716-723.

X. **Wang** & E. **Kerre (1996)**, On the classification and the dependencies of the ordering methods, in: D. Ruan (Ed.), *Fuzzy Logic Foundations and Industrial Applications*, Kluwer Academic Publishers, Dordrecht: 73-88.

X. **Wang** & E.E. **Kerre (2001a)**, Reasonable properties for the ordering of fuzzy quantities (I), *Fuzzy Sets and Systems* 118: 375-385

X. **Wang** & E.E. **Kerre (2001b)**, Reasonable properties for the ordering of fuzzy quantities (II), *Fuzzy Sets and Systems* 118: 387-405.

Y. M. **Wang**, J.B. Yang & D.L. Xu **(2005)**, A preference aggregation method through the estimation of utility intervals *Computers & Operations Research* 32: 2027–2049.

Y. **Yang** & R.I. **John (2003)**, Grey systems and interval valued fuzzy sets. In: *Proceedings of European Society for Fuzzy Logic and Technology- 2003*, Zittau, Germany, 10–12 September: 193–197.

Y.Y. **Yao (1993)**, Interval-set algebra for qualitative knowledge representation. In: *Proceedings of the Fifth International Conference on Computing and Information*: 370–375, (ieeexplore.ieee.org/iel2/923/7608/00315346.pdf).

F. **Zhou**, H.C. Guo, G.X. Chen & G.H. Huang **(2008)**, The Interval Linear Programming: A Revisit, *Journal of Environmental Informatics*, 11 (1): 1-10.

Chapter 3

Acceptability Index and Interval Linear Programming

3.1. Introduction

This chapter defines an interval linear programming problem as an extension of the classical linear programming problem to an inexact environment. Let's refer here a very good example (Tong (1994)) of using interval numbers in an optimization problem:

> There are 1000 chickens raised in a chicken farm and they are raised with two kinds of forages – soya and millet. It is known that each chicken eats 1 – 1.3 kg of forage every day and that for good weight gain it needs at least 0.21 – 0.23 kg of protein and 0.004 – 0.006 kg of calcium everyday. Per kg of soya contains 48 – 52% protein and 0.5 – 0.8% calcium and its price is 0.38 – 0.42 Yuan. Per kg of millet contains 8.5 – 11.5% protein and 0.3% calcium and its price is 0.20 Yuan. How should the forage be mixed in order to minimize expense on forage?

Most of the parameters used in this problem are inexact and perhaps appropriately given in terms of simple intervals. In reality inexactness of this kind can be

A. Sengupta and T.K. Pal: Fuzzy Preference Ordering, STUDFUZZ 238, pp. 39–57.
springerlink.com

cited in countless numbers (Huang et al (1995), Ida (2000), Giove et al (2006), Riverol et al (2006) etc.).

The optimization problem can be structured as follows:

$$\text{Minimize } Z = [0.38,\ 0.42]\, x_1 + 0.20 x_2,$$

subject to

$$x_1 + x_2 = [1,\ 1.3] \times 1000,$$

$$[0.48,\ 0.52]\, x_1 + [0.085,\ 0.115]\, x_2 \geq [0.21,\ 0.23] \times 1000,$$

$$[0.005,\ 0.008]\, x_1 + 0.003\, x_2 \geq [0.004,\ 0.006] \times 1000,$$

$$x_1 \geq 0,\ x_2 \geq 0.$$

However, for solution, the techniques of classical linear programming cannot be applied if and unless the above interval-valued structure of the problem is reduced into a standard linear programming structure (Luhandjula (1986), Rommelfanger (1989), Chakraborty (1995)) and for that we have to clear up the following main issues:

- First, regarding interpretation and realization of the inequality relations involving interval coefficients.
- Second, regarding interpretation and realization of the objective 'Mininimize' with respect to an inexact environment.

In this chapter, we concentrate mainly on a satisfactory solution approach which needs DM's interpretation regarding the inequality relations and the objective of the problem defined in an inexact environment.

This chapter is organized as follows. In Section 3.2, we introduce acceptability index, or $\mathscr{A}$-index in short, for comparing any two interval-numbers. Various properties of this index and an illustrative example have been discussed here. Section 3.3 divided in three subsections gives an elaborate study on the interval inequality relation in order to interpret and realize the relation as an interval-valued constraint of an optimization problem defined in an inexact environment. On the basis of $\mathscr{A}$-index, inequality and equality constraints involving interval coefficients are reduced to their satisfactory crisp equivalent forms. Section 3.4 describes the solution principle of an interval linear programming problem, and discusses efficiency of our methodology in comparison to a previously cited numerical problem (Tong (1994)). Section 3.5 includes the concluding remarks and the future scope.

3.2. The Acceptability Index

Definition 3.2.1. Let $\oslash$ be an extended order relation between the intervals $A=[a_L, a_R]$ and $B=[b_L, b_R]$ on the real line $\Re$, then for $m(A)\leq m(B)$, we construct a premise $A\oslash B$, which implies that A *is inferior to* B (or B *is superior to* A). Here, the term 'inferior to' ('superior to') is analogous to 'less than' ('greater than').

Definition 3.2.2. Let I be the set of all closed intervals on the real line $\Re$. Here, we further define an *acceptability function* $\mathscr{A}: I\times I\rightarrow[0, \infty)$ such that

$$\mathscr{A}(A\oslash B)=\frac{m(B)-m(A)}{w(B)+w(A)},$$

where $w(B)+w(A)\neq 0$. $\mathscr{A}(A\oslash B)$ may be interpreted as *the grade of acceptability* of the '*first interval to be inferior to the second interval*'.

The grade of acceptability of $A\oslash B$ may be classified and interpreted further on the basis of comparative position of mean and width of interval B with respect to those of interval A as follows:

$$\mathscr{A}(A\oslash B)\begin{cases}=0 & \text{if } m(A)=m(B)\\ \in(0,\ 1) & \text{if } m(A)<m(B) \text{ and } a_R>b_L\\ \geq 1 & \text{if } m(A)<m(B) \text{ and } a_R\leq b_L\end{cases}$$

The classification of the acceptability grades are interpreted as follows:

(i) If $\mathscr{A}(A\oslash B)=0$, then the premise '$A$ *is inferior to* B' is not accepted.

(ii) If $0<\mathscr{A}(A\oslash B)<1$, then the interpreter accepts the premise $(A\oslash B)$ with different grades of satisfaction ranging from zero to one (excluding zero and one).

(iii) If $\mathscr{A}(A\oslash B)\geq 1$, the interpreter is absolutely satisfied with the premise $(A\oslash B)$ or, in other words, he accepts that $(A\oslash B)$ is true.

Remark 3.2.3. If $\mathscr{A}(A\oslash B)>0$, then for a maximizing problem (say, A and B are two alternative *interval profits* and the problem is to choose *maximum profit* from among them), interval B is preferred to A and for a minimizing problem (say, A and B are two *interval costs*), A is preferred to B in terms of value.

Remarks 3.2.4. For any sort of value judgment the $\mathscr{A}$-index consistently satisfies the DM: For any two intervals A and B on $\Re$,

$$\text{either, } \mathscr{A}(A \otimes B) > 0$$

$$\text{or, } \mathscr{A}(B \otimes A) > 0$$

$$\text{or, } \mathscr{A}(A \otimes B) = \mathscr{A}(B \otimes A) = 0.$$

Remark 3.2.5. The proposed index is transitive. For any three intervals A, B and C on $\Re$,

$$\text{if } \mathscr{A}(A \otimes B) \geq 0 \text{ and } \mathscr{A}(B \otimes C) \geq 0$$

$$\text{then } \mathscr{A}(A \otimes C) \geq 0.$$

But it *does not mean* that $\mathscr{A}(A \otimes C) \geq \max\{\mathscr{A}(A \otimes B), \mathscr{A}(B \otimes C)\}$.

Proposition 3.2.6. Let $\mathscr{A}(A \otimes B) = \gamma \geq 0$, then for any given interval A, there exists a γ-dependent family of intervals B^{γ} as,

$$B \in \left\{B^{\gamma} = \langle m(B^{\gamma}),\ w(B^{\gamma})\rangle \in \Re, \text{ such that } \mathscr{A}(A \otimes B^{\gamma}) = \gamma\right\}$$

where, mean and width of B^{γ} are linearly dependent to each other, such that

$$m(B^{\gamma}) = \gamma.(w(B^{\gamma})) + k$$

and

$$k = m(A) + \gamma(w(A)).$$

Fig. 3.1 illustrates B^{γ} arrays for a given $A = \langle 15, 5\rangle$ on the *mean-width plane* on which any point is defined as an interval. Here, B^{γ}s are drawn with γ= 0, .2, .4, .6, .8, 1, 1.5, 2, 3 and k is interpreted as B^{γ}s' intercepts on the horizontal axis. The figure reveals a convergence of B^{γ}s at an absurd interval $\langle 15, -5\rangle$ as an interesting observation.

Like B^{γ}, a different family of intervals may be defined as follows:

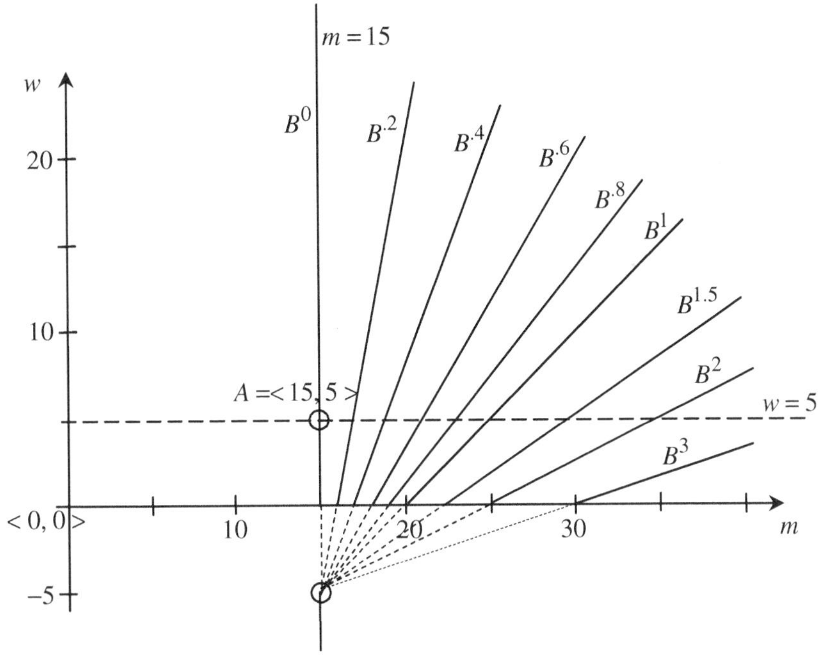

Fig. 3.1. B^{γ} arrays for a given $A = \langle 15, 5 \rangle$ on the mean-width plane

Proposition 3.2.7. Let $\mathscr{A}(A \otimes B) \geq 0$ then there exists a family of intervals

$$\left\{ B_\lambda = \langle m(B_\lambda),\ w(B_\lambda) \rangle : \lambda \text{ is real and } \geq \frac{w(A)}{w(B)+w(A)} \right\} \text{ on } \Re,$$

for which $\mathscr{A}(A \otimes B_\lambda) = \mathscr{A}(A \otimes B)$,
where,

$$m(B_\lambda) = m(A) + \lambda(m(B) - m(A))$$

and

$$w(B_\lambda) = -w(A) + \lambda(w(B) + w(A)).$$

Proof. Let B_1 be an interval such that

$$\mathscr{A}(A \otimes B_1) \ = \ \mathscr{A}(A \otimes B),$$

$$\Rightarrow \frac{m(B_1)-m(A)}{w(B_1)+w(A)} = \frac{m(B)-m(A)}{w(B)+w(A)}$$

$$\Rightarrow \frac{m(B_1)-m(A)}{m(B)-m(A)} = \frac{w(B_1)+w(A)}{w(B)+w(A)} = \lambda, \text{ say,}$$

where, λ is any non-zero finite number. Hence,

$$m(B_1) = m(A) + \lambda(m(B) - m(A))$$

and

$$w(B_1) = -w(A) + \lambda(w(B) + w(A)).$$

Since, $w(B_1)$ cannot be negative, λ is to be restricted by

$$\lambda = \frac{w(A)}{w(B)+w(A)}.$$

Now, as the interval B_1 depends on λ, denoting it by B_λ, we get the theorem.

Remark 3.2.8. If $\lambda_1 < \lambda_2$, then $\mathscr{A}(B_{\lambda_1} \otimes B_{\lambda_2}) > 0$.

Remark 3.2.9. If $\mathscr{A}(A \otimes B) > 0$, then there does not exist any $B' \in \{B_\lambda\}$ for which $\mathscr{A}(A \otimes B') = \mathscr{A}(A \otimes B)$, if

either (i) B' and B are equi-width, i.e., $m(B') \neq m(B)$ and $w(B') = w(B)$

or, (ii) B' and B are equi-centered, i.e., $m(B') = m(B)$ and $w(B') \neq w(B)$.

This implies that through pair-wise comparisons, the degree of acceptability (satisfaction) can be numerically tallied only in a set of intervals where one of the properties – either the central value or the width – remains unchanged.

Remark 3.2.10. If $B_1, B_2, \ldots\ldots, B_n$ be the equi-width intervals on $\Re$, such that

$$\mathscr{A}(B_1 \otimes B_2) = \alpha_1 \geq 0$$

$$\mathscr{A}(B_2 \otimes B_3) = \alpha_2 \geq 0$$

$$\vdots$$

$$\vdots$$

$$\mathscr{A}(B_{n-1} \circleddash B_n) = \alpha_{n-1} \geq 0$$

then

$$\mathscr{A}(B_1 \circleddash B_n) = \alpha_1 + \alpha_2 + \cdots \alpha_{n-1}.$$

Remark 3.2.11. If $\mathscr{A}(B_1 \circleddash B_2) = 0$ and $w(B_1) = w(B_2)$, then $B_1 \equiv B_2$, i.e., B_1 is identical to B_2.

Now, if $\mathscr{A}(B_1 \circleddash B_2) = 0$ and $w(B_1) \neq w(B_2)$, then obviously B_1 *is not identical to* B_2 (or, symbolically, $B_1 \not\equiv B_2$). But through $\mathscr{A}$-index, direct comparisons between them interpret that B_1 and B_2 are non-inferior to each other. Then the question arises: *How to choose a preferred (maximizing) alternative?*

Proposition 3.2.12. Let us consider an interval $D = \langle m(D),\ w(D)\rangle$ which is inferior to the equi-centered intervals B_1 and B_2. Then, as compared to *D*, the grade of acceptability of superiority of the less uncertain interval is higher than that of superiority of the more uncertain interval. Symbolically, if

$$\mathscr{A}(D \circleddash B_1) > 0, \quad \mathscr{A}(D \circleddash B_2) > 0$$

and

$$\mathscr{A}(B_1 \circleddash B_2) = 0, \text{ but } B_1 \not\equiv B_2,$$

then

(i) $\mathscr{A}(D \circleddash B_1) > \mathscr{A}(D \circleddash B_2)$ *iff* $w(B_1) < w(B_2)$,

(ii) $\mathscr{A}(D \circleddash B_1) < \mathscr{A}(D \circleddash B_2)$ *iff* $w(B_1) > w(B_2)$.

Proof. The proof is straight-forward.

Here condition (i) indicates that as compared to *D*, superiority of B_1 is more believable than the superiority of B_2. Hence, B_1 must be preferred to B_2 as maximizing alternative. This result is quite compatible to our intuition: if intervals B_1 and B_2 have the same expected value but B_1 contains less uncertainty than B_2 then B_1 is preferred to B_2.

Now, what happens if the reference interval $D^* = \langle m(D^*),\ w(D^*)\rangle$ is taken to be superior to the equi-centered intervals B_1 and B_2, i.e., if

$$\mathscr{A}(B_1 \circleddash D^*) > 0, \quad \mathscr{A}(B_2 \circleddash D^*) > 0$$

and

$$\mathscr{A}(B_1 \circleddash B_2) = 0, \text{ but } B_1 \not\equiv B_2,$$

then

(i) $\mathscr{A}(B_1 \circleddash D^*) > \mathscr{A}(B_2 \circleddash D^*)$ *iff* $w(B_1) < w(B_2)$,

(ii) $\mathscr{A}(B_1 \circleddash D^*) < \mathscr{A}(B_2 \circleddash D^*)$ *iff* $w(B_1) > w(B_2)$.

Here condition (i) states that as compared to D^*, inferiority of B_1 is more believable than the inferiority of B_2. Therefore, B_1 must be preferred to B_2 as better minimizing alternative. On the contrary, from this condition, B_2 cannot be said to be preferred to B_1 as maximizing alternative.

Here is an important point to be noted: The concept of acceptability index for comparing intervals in no way can be treated as analogous to the concept of difference in the real analysis. And for this reason, considering a superior reference interval D^* for choosing a preferred maximizing alternative from among the equi-centered but not identical B_1 and B_2 or, considering an inferior reference D for choosing a preferred minimizing alternative from among B_1 and B_2 do not make any sense and yield nothing.

3.2.1. *Illustrative example*

Example 3.2.1.1. Let A and B_1 be two alternative profit intervals as such

$$A = [120,\ 180] = \langle 150,\ 30 \rangle$$

$$B_1 = [130,\ 210] = \langle 170,\ 40 \rangle$$

and the problem is to find the maximizing alternative.

$$\mathscr{A}(A \circleddash B_1) = 0.28.$$

Therefore, the DM is satisfied that A is inferior to B_1 and his grade of satisfaction is 0.28. B_1 is the maximizing alternative.

Example 3.2.1.2. Now let B_1 in *Example 3.2.1.1* is replaced by B_2, where

$$B_2 = [120,\ 220] = \langle 170,\ 50 \rangle,$$

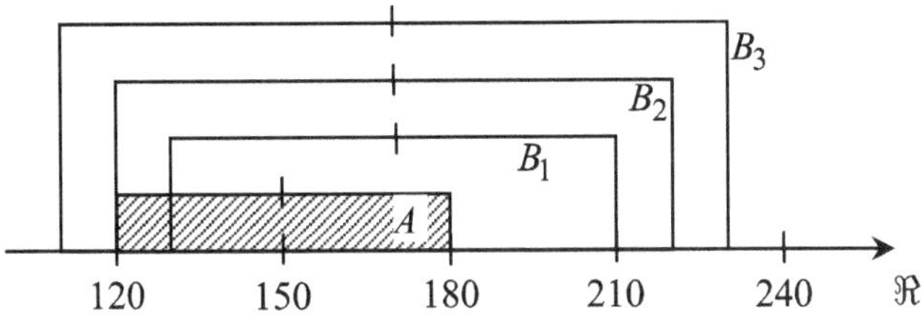

Fig. 3.2: Graphical representation of example 3.2.1.

(*note that,* $m(B_2) = m(B_1)$ but $w(B_2) > w(B_1)$). Now, to choose the maximizing profit interval from among A and B_2. Then

$$\mathscr{A}(A \circledcirc B_2) = 0.25.$$

$\Rightarrow B_2$ is the maximizing alternative.

It also tells that the DM is satisfied with B_2's superiority over A with the grade of satisfaction at 0.25 (which is lower than the satisfaction grade obtained in *Example 3.2.1.1.*)

Although B_2 and B_1 are equi-centered, it is only the B_2's higher uncertainty that makes the difference in the grade of satisfaction for *Examples 3.2.1.1* and *3.2.1.2.*

Remark 3.2.1.3. According to Ishibuchi & Tanaka (1990) , the order relation $\leq_{LR}$ holds for both of the examples (hence, $\leq_{mw}$ is not applicable). Same results are also obtained by using $\leq_{LR}$ but it fails to interpret further (about DM's satisfaction over the selections) because the issue of uncertainty is not taken in consideration.

Example 3.2.1.4. Let the extent of uncertainty in B_2 of *Example 3.2.1.2* be increased further with same mean. Let $B_3 = [110, 230] = \langle 170, 60 \rangle$. To find maximizing alternative from among A and B_3.

Remark 3.2.1.5. This is a particular case of nested interval; A is nested in B_3. Here both of the order relations $\leq_{LR}$ and $\leq_{mw}$ fails (Ishibuchi & Tanaka (1990)). But the proposed $\mathscr{A}$-index may shed some light on such problems. We have

$$\mathscr{A}(A \circledcirc B_3) = 0.22.$$

$\Rightarrow B_3$ is the maximizing, though DM's satisfaction (grade of acceptability) is reduced further.

3.3. A satisfactory crisp equivalent system of $Ax \leq B$

3.3.1. *Tong's approach*

Tong (1994) deals with interval inequality relations in a separate way. For a minimization problem as follows,

$$\text{Minimize } Z = \sum_{j=1}^{n} [c_{Lj}, c_{Rj}]x_j,$$

Subject to,

$$\sum_{j=1}^{n} [a_{Lij}, a_{Rij}]x_j \geq [b_{Li}, b_{Ri}], \quad \forall i = 1, 2, \ldots, m,$$

$$x_j \geq 0, \quad \forall j,$$

each inequality constraint is first transformed into 2^{n+1} crisp inequalities to yield

$$D_i = \left\{ D_i^k / k = 1, 2, \ldots\ldots, 2^{n+1} \right\},$$

which are the solutions to the *i*th set of 2^{n+1} inequalities.

On the other hand, Tong defines a characteristic formula (CF),

$$\sum_{j=1}^{n} a_{ij}x_j \geq b_j$$

of the *i*th inequality relation, $\forall i$, where $a_{ij} \in [a_{Lij}, a_{Rij}]$ and $b_i \in [b_{Li}, b_{Ri}]$.

Now, if the ith CF generates solution D_i such that

$$\overline{D_i} = \bigcup_{k=1}^{2^{n+1}} D_i^k,$$

then CF is called *maximum-value range* inequality and if CF generates solution D_i, such that

$$\underline{D_i} = \bigcap_{k=1}^{2^{n+1}} D_i^k,$$

then it is called *minimum-value range* inequality.

Tong (1994) then defines minimum and maximum optimal objective value of the problem using max and min value inequalities, respectively.

3.3.2. *Discussion*

Let us take a very simple inequality relation with a single variable,

$$[10,\ 20]x \leq [5,\ 35].$$

According to Tong (1994)'s approach, the interval inequality generates 2^{1+1} crisp inequalities:

$$\left.\begin{array}{l} 10x \leq 5 \Rightarrow x \leq 0.5 \\ 10x \leq 35 \Rightarrow x \leq 3.5 \\ 20x \leq 5 \Rightarrow x \leq 0.25 \\ 20x \leq 35 \Rightarrow x \leq 1.75 \end{array}\right\} D = \left\{D^k / k = 1,\ 2,\ 3,\ 4\right\}$$

$$\overline{D} = \bigcup_{k=1}^{2^2} D^k \Rightarrow x \leq 3.5 : \textit{max-value range} \text{ inequalities.}$$

$$\underline{D} = \bigcap_{k=1}^{2^2} D^k \Rightarrow x \leq 0.25 : \textit{min-value range} \text{ inequalities.}$$

Here we would like to raise a question on Tong (1994)'s approach: how does one interpret the use of the operators *union* and *intersection* in defining max- and min-value range inequalities, respectively?

Using the union operator in defining the crisp equivalent form of the *i*th original constraint may be interpreted as at least one element of the interval $A_i x$ is less than or equal to at least one element of interval B_i which clearly does not validate the original constraint condition. Using $\mathscr{A}$-index it can be shown that

$$\mathscr{A}(B_i \odot A_i x) = 1,$$

i.e., $A_i x$ is definitely greater than B_i.

On the other hand, using the intersection operator in defining the crisp equivalent form may be interpreted as *all elements of* $A_i x$ *is less than or equal to all elements of* B_i which is merely an oversimplification. Using $\mathscr{A}$-index, it can be shown that

$$\mathscr{A}(A_i x \ominus B_i) = 1 .$$

In actual practice, for a wide range of feasibility of the decision variable vector, DM may allow $A_i x$ even to be nested in B_i, i.e., some/all elements of $A_i x$ may even be allowed to be greater than or equal to some elements of B_i and that how much to be allowed will be decided by the DM and this will depend on his optimistic attitude, on his risk versus benefit assessment and as a whole, on the level of satisfaction the DM tries to achieve from the decision-making process.

Hence, in our opinion, some sort of conditions indicating DM's satisfaction/utility requirement has to be incorporated in generation of a crisp equivalent structure of the inequality constraint with interval coefficients. Using the properties of $\mathscr{A}$-index we develop a satisfactory crisp equivalent structure of an inequality constraint with interval coefficients.

3.3.3. *A satisfactory crisp equivalent system of* $Ax \le B$ *based on* $\mathscr{A}$*-index*

Let $A = [a_L, a_R]$, $B = [b_L, b_R]$ and x is a singleton variable.

According to $\mathscr{A}$-index the acceptability condition of $Ax \le B$ may be defined as $\mathscr{A}(Ax \ominus B) \ge 0$, i.e.,

$$m(Ax) \le m(B) .$$

Now, let us take the condition $m(Ax) = m(B)$, then, for a given value of x, we may have two different possible set-ups.

Case-I: When interval A is relatively narrower than interval B:
Ax may be nested in B. For example, for $x = 2$, the relation $[2, 4]x \le [2, 10]$ may be viewed as given in Fig. 3.3.

Case-II: When interval A is relatively wider than as it was in case I:
B may be nested in Ax. For example, for $x = 2$, the relation $[0, 6]x \le [2, 10]$ may be viewed as shown in Fig. 3.4.

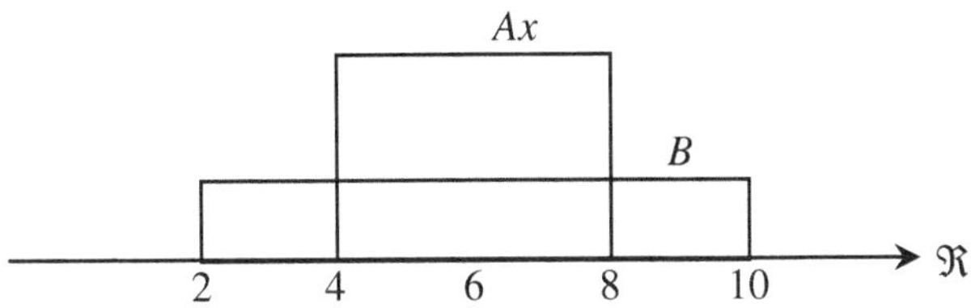

Fig. 3.3. Interval Ax is nested in interval B.

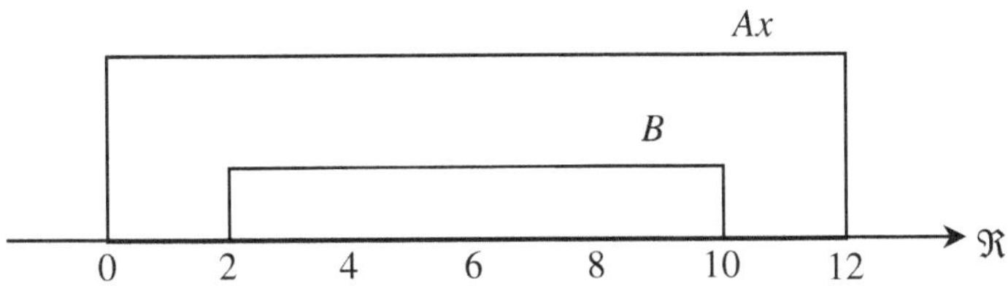

Fig. 3.4. Interval B is nested in interval Ax.

From the examples given above, for both of the cases, the following remarks can be made:

(i) Case I definitely satisfies the original interval inequality for $x = 2$ because $\mathscr{A}(Ax \ominus B) \geq 0$. However, an optimistic DM may remain under-satisfied with the optimal constraint condition and for getting higher satisfaction, he may like to increase the value of x to such an extent that $\mathscr{A}(B \ominus Ax)$ does not pass over a threshold assumed and fixed by him.

(ii) On the other hand, by case II, the original interval inequality condition is not denied even for $x \leq 2$ because $\mathscr{A}(Ax \ominus B) \geq 0$. But a pessimistic DM may not be satisfied if the right limit of Ax spills over the right limit of B. To attain his required level of satisfaction the DM may even like to reduce the value of x so that $a_R\, x \leq b_R$.

To shed more light on interval inequality relation from a different angle, let us refer an equivalent form of a deterministic inequality where

$$ax \leq b \text{ is } ax \in\ ]-\infty,\ b]\,.$$

Let us extend this concept to an inexact environment: if the real numbers a and b are allowed to be replaced by intervals A and B, respectively, that one's possible reaction is as much as similar to Moore's concept of set-inclusion, i.e.,

$$Ax \leq B \Rightarrow Ax \subset D \text{, where, } D = \;]-\infty,\ b_R\,].$$

Keeping in view the two remarks stated above and the Moore's concept (1979), we propose a satisfactory crisp equivalent form of interval inequality relation as follows:

$$Ax \leq B \;\Rightarrow\; \begin{cases} a_R\, x \leq b_R \\ \mathscr{A}(B \ominus Ax) \geq \alpha \in [0,\ 1]. \end{cases}$$

where, α may be interpreted as an optimistic threshold assumed and fixed by the DM.

Similarly, for $Ax \geq B$, we have the satisfactory crisp equivalent form by the following pair:

$$\begin{cases} a_L\, x \geq b_L\,, \\ \mathscr{A}(Ax \ominus B) \leq \alpha \in [0,\ 1]. \end{cases}$$

3.4. An Interval Linear Programming Problem and its Solution

Let us consider the following problem:

$$\text{Minimize } \; Z = \sum_{j=1}^{n} [c_{Lj},\ c_{Rj}]x_j$$

subject to,

$$\sum_{j=1}^{n} [a_{Lij},\ a_{Rij}]x_j \geq [b_{Li},\ b_{Ri}], \quad \forall i = 1, 2, \ldots., m.$$

$$x_j \geq 0, \quad \forall j.$$

As is described in the previous section a satisfactory crisp equivalent system of constraints of the ith interval constraint can be generated as follows:

$$\sum_{j=1}^{n} a_{Lij}x_j \geq b_{Li}, \quad \forall i,$$

$$b_{Li} + b_{Ri} - \sum_{j=1}^{n} (a_{Lij} + a_{Rij})x_j \leq \alpha(b_{Ri} - b_{Li}) + \alpha \sum_{j=1}^{n} (a_{Rij} - a_{Lij})x_j.$$

The working of $\mathscr{A}$-index may be summarized by the following principle:

> The position (of mean) of an interval compared to that of another reference interval results in whether the former is superior or inferior to the latter. On the other hand, the width of a superior (inferior) interval compared to that of the reference interval specifies the grade to which the DM is satisfied with the superiority (inferiority) of the former compared to the latter.

The purpose of a conventional linear programming problem is to maximize or minimize the value of its (one only, single-valued) objective function satisfying a given set of restrictions. But, a single-objective interval linear programming problem contains an interval-valued objective function. As an interval can be represented by any two of its four attributes (viz., left limit, right limit, mid-value and width (Ishibuchi & Tanaka (1990)), an interval linear programming, by using attributes mid-value and width (say) can be reduced into a linear bi-objective programming problem as follows:

Max/Min {mid-value of the interval objective function},

Min {width of the interval objective function},

subject to,

{set of feasibility constraints}.

From this problem naturally one may get two conflicting optimal solutions:

$$x^* = \{x_j^*\}, \text{ from } \begin{cases} \text{Max/Min } \{\text{mid value}\} \\ \text{sub. to, } \{\text{constraints}\}, \end{cases}$$

$$x^{**} = \{x_j^{**}\}, \text{ from } \begin{cases} \text{Min } \{\text{width}\} \\ \text{sub. to, } \{\text{constraints}\}, \end{cases}$$

and hence we get two optimal interval values Z^* and Z^{**}.

If $x^* = x^{**}$, then there does not exist any conflict and x^* is the solution of the problem.

But if $x^* \neq x^{**}$, for the maximization problem, we do have the condition as follows,

$$m(Z^*) > m(Z^{**}) \text{ and } w(Z^*) > w(Z^{**})$$

(because, Z^* is obtained through maximizing $m(Z)$ and Z^{**} is obtained not by maximizing $m(Z)$, but through another goal, by minimizing $w(Z)$).

Similarly, for a minimizing problem, if $x^* \neq x^{**}$, then, we have,

$$m(Z^*) < m(Z^{**}) \text{ and } w(Z^*) > w(Z^{**})$$

(because, Z^* here is obtained through minimizing $m(Z)$ and Z^{**} by minimizing $w(Z)$).

Therefore, if $x^* \neq x^{**}$, Z^* and Z^{**} are said to be two non-dominated alternative extreme interval objective values (Ignizio (1982)).

On the other hand, the principle of $\mathscr{A}$-index indicates that for the maximization (minimization) problem, an interval with a higher mid-value is superior (inferior) to an interval with a lower mid-value.

Though Z^* and Z^{**} are two non-dominated alternatives from the viewpoint of a bi-objective problem, we like to see it differently. These intervals may also be viewed as two functional values of the interval-valued objective function of the original problem. So, they can be ranked through $\mathscr{A}$-index. Hence, in order to obtain max/min of the interval objective function, considering the mid-value of the interval-valued objective function is our primary concern. We reduce the interval objective function into its central value and use conventional LP techniques for solution. We also consider width but as a secondary attribute, only to confirm whether it is within the acceptable limit of the DM. If it is not, one has to reduce the extent of width (uncertainty) according to his satisfaction and thus to obtain a less wide interval from among the non-dominated alternatives accordingly.

The following LP problem is the necessary equivalent form of the original problem:

$$\text{Minimize } m(Z) = \tfrac{1}{2}\sum_{j=1}^{n}(c_{\mathrm{L}j} + c_{\mathrm{R}j})x_j$$

subject to,

$$\sum_{j=1}^{n} a_{\mathrm{L}ij}x_j \geq b_{\mathrm{L}i}, \quad \forall i,$$

$$b_{\mathrm{L}i} + b_{\mathrm{R}i} - \sum_{j=1}^{n}(a_{\mathrm{L}ij} + a_{\mathrm{R}ij})x_j \leq \alpha(b_{\mathrm{R}i} - b_{\mathrm{L}i}) + \alpha\sum_{j=1}^{n}(a_{\mathrm{R}ij} - a_{\mathrm{L}ij})x_j,$$

$$x_j \geq 0, \quad \forall j.$$

It is only when the possibility of multiple solutions arises, then only the comparative widths are required to be calculated and consequently in favour of a minimum available width, we get the solution.

3.4.1. *Solution to the problem stated in Section 3.1.*

Here we transform the numerical problem stated in Section 3.1 in its crisp equivalent form as follows:

$$\text{Minimize } 0.4x_1 + 0.2x_2$$

subject to

$$x_1 + x_2 \geq 1000$$

$$x_1 + x_2 \leq 1300$$

$$0.48x_1 + 0.085x_2 \leq 210$$

$$0.51x_1 + 1.0075x_2 \leq 215$$

$$5x_1 + 3x_2 \leq 4000$$

$$8x_1 + 3x_2 \leq 4500$$

$$x_1, x_2 \geq 0$$

(The DM here assumes that $\alpha = 0.5$).

Result. $x_1 = 285.71$,

$x_2 = 857.14$ and

$Z_{0.5} = [279.99,\ 291.43] = \langle 285.71,\ 5.72\rangle$

(where, $Z_{0.5}$ is the minimum objective value at $\alpha = 0.5$).

Tong's approach gives the following solution for the same problem:

$$x_1 = [x_1', x_1''] = [234.57,\ 1050],$$

$$x_2 = [x_2', x_2''] = [765.43,\ 250]$$

and

$$Z_{\text{Tong}} = [242.22,\ 491] = \langle 366.61,\ 124.39\rangle .$$

As far as the problem's minimizing objective is concerned, our solution gives better expected value with far better certainty. Also by using $\mathscr{A}$-index, we have,

$$\mathscr{A}(Z_{0.5} \oslash Z_{\text{Tong}}) = 0.62.$$

Clearly, this indicates that the solution with our approach is much better than that with Tong's approach. Even if we compute the satisfying conditions of the crisp equivalent set of constraints at $\alpha = 0$, we get the solution to the above problem as

$$x_1 = 571.28$$

$$x_2 = 428.57$$

and

$$Z_0 = [302.85,\ \ 325.71] = \langle 314.28,\ \ 11.42 \rangle$$

for which also, we have,

$$\mathscr{A}(Z_0 \oslash Z_{\text{Tong}}) = 0.385.$$

This indicates that even at $\alpha = 0$ (at a no optimism level) our solution is better than Tong's solution.

3.5. Conclusion

The aim of this chapter was to define a satisfactory crisp equivalent system of an inequality constraint with interval coefficients. The approach defined here has come out as an application of $\mathscr{A}$-index for comparing two intervals through DM's satisfaction. Once the crisp equivalent structure of the constraint set is defined, solution to a problem with maximizing or minimizing objective function practically turns to be maximization or minimization of the central value of the interval-valued objective function. In this regard, a point worth mentioning is: If the DM is not satisfied with the extent of uncertainty (width) involved in the optimal objective value, he can achieve his required level of satisfaction by adjusting allowable width of the optimal objective value and/or by redefining satisfying conditions for generating crisp equivalent set of constraints. An interactive approach, through which a DM learns gradually more and more about the problem and where his feedback is used to guide the solution of the problem to a more favourable one, may be formulated as a generalized and more credible procedure for solving interval linear programming problems.

3.6. References

Debjani **Chakraborty (1995)**, *Optimization in Imprecise and Uncertain Environment*, Ph. D. Thesis, Dept. of Mathematics, I.I.T. Kharagpur.

S. **Giove**, S. Funari & C. Nardelli (**2006**), An interval portfolio selection problem based on regret function, *European Journal of Operational Research* 170: 253–264.

G.H. **Huang**, B.W. Baetz & G.G. Patry (**1995**), Grey integer programming: An application to waste management planning under uncertainty, *European Journal of Operational Research* 83: 594-620.

M. **Ida (2000)**, Interval multiobjective programming and mobile robot path planning, In: M. Mohammadian (Ed.): *New Frontier in Computational Intelligence and its Applications*, IOS Press: 313-322.

J.P. **Ignizio (1982)**, *Linear Programming in Single and Multiple Objective Systems*, Prentice-Hall, Englewoods Cliffs.

H. **Ishibuchi** & H. **Tanaka (1990)**, Multiobjective programming in optimization of the interval objective function, *European Journal of Operational Research* 48 (2): 219 –225.

M.K. **Luhandjula (1986)**, On Possibilistic Linear Programming, *Fuzzy Sets and Systems* 18: 15-30.

R.E. **Moore (1979)**, *Method and Application of Interval Analysis*, SIAM, Philadelphia.

C. **Riverol**, M.V. Pilipovik & C. Carosi (**2006**), Assessing the water requirements in refineries using possibilistic programming, *Chemical Engineering and Processing* 45 (7): 533-537.

H. **Rommelfanger (1989)**, Interactive decision making in fuzzy linear optimization problems, *European Journal of Operational Research* 4: 210-217.

S. **Tong (1994)**, Interval number and fuzzy number linear programming, *Fuzzy Sets and Systems* 66: 301–306.

Chapter 4

Fuzzy Preference Ordering of Intervals

4.1. Introduction

In the formulation of realistic problems, set of intervals may appear as coefficients in the inequality (or equality) constraints of an optimization problem or they may appear in the selection of a preferred alternative from among a set of many in a decision making problem. Consequently, there might be a key question related to the comparison of any two interval-numbers.

Moore (1966, 1979)'s pioneering works on the arithmetic of interval numbers provide two transitive order relations defined over intervals, one as an extension of '$<$' on the real line and another as an extension of '$\subseteq$' as the concept of set inclusion. But these order relations cannot explain ranking between two partially or fully overlapping intervals. Ishibuchi & Tanaka (1990)'s ranking methodology, which is considerably advanced over Moore (1979), suggested two order relations $\leq_{LR}$ and $\leq_{mw}$ in *either or survivor* basis. However, there exist a set of pair of intervals for which both the order relations do not hold. Moreover, these order relations did not discuss anything on *how much higher* an interval when it is known to be higher than another. If it is argued that the interval numbers are generated through the inexactness of the problem environment, there should be an attempt to

A. Sengupta and T.K. Pal: Fuzzy Preference Ordering, STUDFUZZ 238, pp. 59–89.
springerlink.com

satisfy (most likely subjectively) the query 'how much higher one interval is than another?' From this viewpoint Kundu (1997)'s *fuzzy leftness relationship* is more reasonable but for a particular instance of nested intervals it produces an inconsistent conclusion. A detail discussion on comparing different approaches has been given in Chapter 2.

In order to answer the 'whys and hows' raised in this discussion we have introduced $\mathscr{A}$-index for comparing any two interval-numbers in Chapter 3. $\mathscr{A}$-index defines an objective way to compare and order any two interval-numbers in terms of value. $\mathscr{A}$-index serves better if it is used as a value judgment index rather than if it is used as a rational human being's choice or preference index. Here in this chapter in Section 4.2, the strength and weakness of $\mathscr{A}$-index over the other approaches have been studied. Next, in Section 4.3, to supplement the value-based approach, we define the second approach as a preference approach of ranking two interval-numbers from a pessimistic (or risk averse) DM's point of view. In Section 4.4, we have generalized the preference index to be applicable under different degree of pessimism (or optimism). Here we have defined a DM's final choice in linguistic term and given some numerical examples. Fuzzy preference ordering is then discussed for minimizing objective of the DM in Section 4.5. The final section gives a study in detail on the comparative advantage of the new index over the other indices. This section also includes a brief discussion on some latest indices appeared after 2000 – 01 in the literature of interval ranking.

4.2. The strength and weakness of $\mathscr{A}$-index

A general statement on the working of $\mathscr{A}$-index may be made: The position (of mean) of an interval compared to that of another reference interval results in whether the former is superior or inferior to the latter. On the other hand, the width of a superior (or inferior) interval compared to that of the reference interval specifies the grade to which the DM is satisfied with the superiority of (inferiority) of the former compared to the latter.

Now a natural question comes: Is it necessary to study (the value of) the proposed index? Our answer is – Yes. Primarily, an $\mathscr{A}$-index value interprets that whether an interval is superior (or inferior) to another or not. Moreover, in a real decision situation mere knowing of whether an interval is superior (or inferior) to another is not sufficient, even an optimistic DM must be fond of knowing with which grade of satisfaction he is taking a decision.

$\mathscr{A}$-index is applicable to any pair of intervals on the real line whereas the approach of Ishibuchi & Tanaka (1990) is limited when both of the relations $\leq_{LR}$ and $\leq_{mw}$ are inapplicable. To compare any two intervals on the real line the *fuzzy leftness relationship*, introduced by Kundu (1997), obtains almost same conclusion (as $\mathscr{A}$-index yields) except one case. For any two intervals A and B, if $m(A) = m(B)$ and $w(A) \neq w(B)$, Kundu's approach indicates that both A and B are optimal choice which is clearly inconsistent to our intuition. On this particular point, $\leq_{\text{mw}}$ of Ishibuchi & Tanaka (1990) and $\mathscr{A}$-index lead to the same conclusion that the less uncertain interval will be the *best choice* for both of the maximization and minimization problems. Apart from simplicity of calculations, goodness of $\mathscr{A}$-index perhaps lies in the success of the Remark 3.2.9 (and hence in the success of Remark 3.2.10 and Proposition 3.2.12).

Another advantage of $\mathscr{A}$-index is that it can be utilized to interpret and define the satisfactory equivalent deterministic set of constraints of one interval-valued inequality constraint,

$$\sum A_i x_i \leq B \qquad \text{or} \qquad \sum A_i x_i \geq B,$$

in a mathematical programming problem, where $x = \{x_1, x_2, ..., x_n\}$ is the decision variable vector and $A_1, A_2, ..., A_n, B$ are the interval numbers (Section 3.3). As against the simplicity of $\mathscr{A}$-index, on this point the main disadvantage of using fuzzy leftness relationship (Kundu (1997)) will be the complexity of calculation due to the presence of decision variable x in the construction of membership function defined by conditional probabilities.

The main weakness of $\mathscr{A}$-index is that, as it is only a value-based ranking index, only partially it can be applicable for selection of the best alternative from a pessimistic (or a risk averse) DM's point of view. In another words, only an optimistic DM can use it completely.

For any pair of intervals, apart from the value of $\mathscr{A}$-index, for selection of his choiced alternative, a pessimistic DM will search for another assurance. To complete the applicability of $\mathscr{A}$-index as a pessimistic DM's preference index too, we introduce a supplementary preference relation for a class of intervals.

4.3. Preference ordering for a pessimistic DM for a maximization problem

In order to develop specific preference pattern between a pair of intervals on the real line from a pessimistic DM's point of view, first let us assume that A and B are two interval profits and the problem is to choose maximum profit (interval) from among them. Here, the term *maximum profit* may be interpreted as the best profit from a pessimistic (or, risk averse) DM's point of view.

***Assumptions 4.3.1*:** To guide the analysis, assumptions for the maximization problem are as follows:

1. More money is better than less money.
2. More certainty is better than less certainty.

3. If more money is associated with more uncertainty, a DM undergoes a trade-off between the two.
4. To a pessimistic DM, assumption 2 is somewhat more important than assumption 1. (Obviously, to an optimistic DM, assumption 1 is somewhat more important than assumption 2.)

Definition 4.3.2. Any pair of intervals can be classified into two sets, $\mathcal{S}_1$ and $\mathcal{S}_2$. If $\mathscr{A}(A \oslash B)$ is constructed for A and B then the pair

(i) $(A, B) \in \mathcal{S}_1$, if $\mathscr{A}(A \oslash B) \geq 0$ and $w(A) \geq w(B)$,

(ii) $(A, B) \in \mathcal{S}_2$, if $\mathscr{A}(A \oslash B) \geq 0$ and $w(A) < w(B)$.

For a maximization problem for $(A, B) \in \mathcal{S}_1$, unless A and B are identical, B is always the best choice, i.e., B is strictly preferred to A. This inference can be made simply on the basis of assumptions 1 and 2. Moreover, $\mathcal{S}_1$ satisfies all the conditions to fulfil $\leq_{\text{mw}}$ relation of Ishibuchi & Tanaka (1990)'s approach.

However, this type of crisp inference cannot be made for $(A, B) \in \mathcal{S}_2$ as far as a pessimistic DM's choice is concerned. Depending on the situation, interval B may be strictly preferred to A, A may be strictly preferred to B or there may be a fuzzy preference between A and B.

For any pair (A, B) in $\mathcal{S}_2$,

if $\mathscr{A}(A \oslash B) \geq 1$, then B is strictly preferred to A,

or,

if $\mathscr{A}(A \oslash B) = 0$, then A is strictly preferred to B,

else

if $\mathscr{A}(A \oslash B) \in]0, 1[$, then there exists a *fuzzy preference* between A and B.

Now, in order to develop fuzzy preference between the pair (A, B) in $\mathcal{S}_2$, let us define a fuzzy set, B' as *rejection of B* in $\mathcal{S}_2$. To define its membership function $\mu_{B'}$, we restate the formation of $\mathcal{S}_2$ for the sake of clarity of fuzzy set theoretic logic as well as for notational simplicity.

Definition 4.3.3. For a given interval $B = [b_L, b_R] = \langle m(B), w(B) \rangle$, there exists a set of intervals (X, B) in $\mathcal{S}_2$, where $X = [x_L, x_R] = \langle m(X), w(X) \rangle$ is assumed to be a variable interval with changing mean $m(X)$ and width $w(X)$, remaining all other properties of $\mathcal{S}_2$ unchanged. Hence, we define the fuzzy set B' as the *rejection of B* compared to X,

$$B' = \{(X,\ B) \in \mathcal{S}_2 \ / \ \mathscr{A}(X \circledcirc B) \geq 0,\ \ w(X) < w(B)\}$$

with the membership function $\mu_{B'}(X,B)$, $\mu_{B'}$ being a function $\mathcal{S}_2 \to [0,\ 1]$ (Fig. 4.1),

$$\mu_{B'}(X,B) = \begin{cases} 1 & \text{if } m(X) = m(B), \\ \max\left\{0, \dfrac{m(X) - (b_L + w(X))}{m(B) - (b_L + w(X))}\right\} & \text{if } m(B) \geq m(X) \geq b_L + w(X), \\ 0 & \text{otherwise.} \end{cases}$$

Here, we can interpret different values of $\mu_{B'}$ as follows:

if $\mu_{B'} = 1$, then B is definitely rejected;

if $\mu_{B'} = 0$, then B is definitely accepted and

if $\mu_{B'}$ is in between 0 and 1; then this indicates different degrees of *rejection of B*.

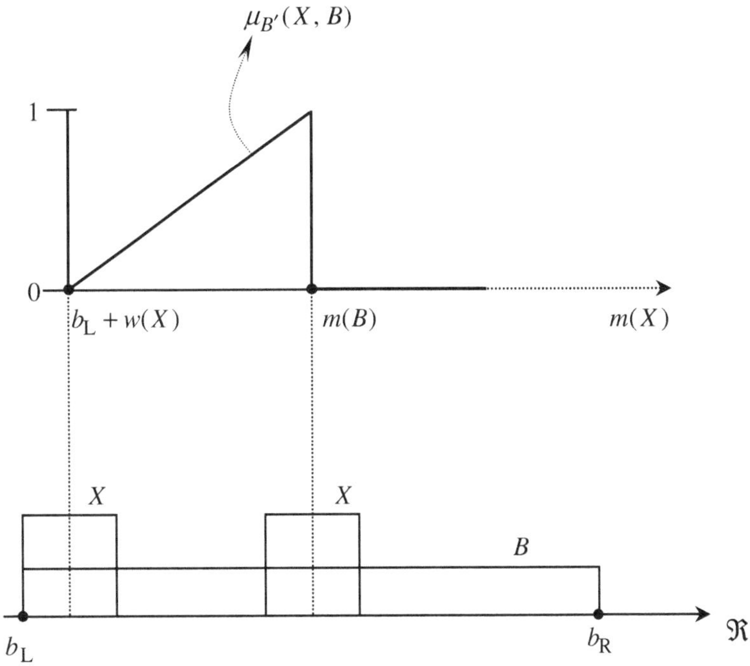

Fig. 4.1. Membership function of *rejection of B* for the pair of intervals (X,B) in S_2.

For consistency and completeness in the formulation of preference structure for any pair (X, B) in $\mathcal{S}_2$, let us define another fuzzy set, X', *rejection of X* in $\mathcal{S}_2$, as the complement of B', i.e.,

$$\forall X \text{ of } (X, B) \in \mathcal{S}_2, \ \mu_{X'}(X,B) = 1 - \mu_{B'}(X,B).$$

Therefore, for any pair of intervals (X, B) in $\mathcal{S}_2$, the pair of fuzzy sets (X', B') may be called a *fuzzy partition* between their preferences (Lowen (1978)). In other words, in search of a *satisfactory maximizing decision* (Bellman & Zadeh (1970), Zimmermann (1991) and Yen & Langari (1999)) for any pair (X, B) in $\mathcal{S}_2$, $\mu_{X'}$ describes the degree of dominance of B over X and $\mu_{B'}$ describes the degree of dominance of X over B.

So, there exists a fuzzy set of non-dominated preferences, *nond* for some pair of intervals (X, B) in $\mathcal{S}_2$ and is defined as

$$nond = X' \cap B',$$

where, $\mu_{\underset{(X,B) \in \mathcal{S}_2}{nond}}(X,B) = \min\,\{\mu_{X'}(X,B),\ \mu_{B'}(X,B)\}.$

A DM hence may choose the pairs of intervals (X, B) in $\mathcal{S}_2$, as the pairs of non-dominated preferences,

$$\text{if } \mu_{nond} \text{ is close enough to } \sup_{(X,\,B) \in \mathcal{S}_2} \mu_{nond}.$$

If the DM is not interested in a fuzzy set of non-dominated alternatives but in a crisp set of maximal non-dominated pair of intervals, (X^0, B) in $\mathcal{S}_2$, then

$$\mu_{nond}(X^0, B) = \sup_{(X,\,B) \in \mathcal{S}_2} \mu_{nond}(X,B).$$

4.4. Choice of the DMs with different degrees of pessimism

To define fuzzy preference ordering for the pair of intervals (X, B) in $\mathcal{S}_2$ for different DMs with different degrees of pessimism (the degrees of pessimism may also be interpreted as the intensities of risk preferences, i.e., high degree of pessimism may be interpreted as low intensity of risk preference and low degree of pessimism as high intensity of risk preference), we modify the construction of B'.

We define B' as *rejection of B for a DM with a specific degree of pessimism* with the non-linear membership function,

$$\pi_{B'} = (\mu_{B'})^p, \quad \mu_{B'} \in [0, 1],$$

where, p is the (non-fuzzy) level of pessimism parameter for any DM and numerically $p = [(\frac{1}{M}), M]$, where M is a finite large number. When the value of the pessimism parameter, $p = \frac{1}{M}$, the DM is said to be of an absolutely pessimistic person who practically does not take any risk. His grade of rejection of B for any (X, B) in $\mathcal{S}_2$ may be explained as

$$\pi_{B'}^{\text{Absolutely pessimistic}} = (\mu_{B'})^{\frac{1}{M}} = \begin{cases} 0 & \text{for } m(X) \le b_L + w(X), \\ 1 & \text{for } m(X) > b_L + w(X). \end{cases}$$

Similarly, for an absolutely optimistic DM, we use $p = M$ and get the membership function for rejection of B as

$$\pi_{B'}^{\text{Absolutely optimistic}} = (\mu_{B'})^{M} = \begin{cases} 1 & \text{for } m(X) = m(B), \\ 0 & \text{for } m(X) < m(B), \end{cases}$$

which is actually the conclusion of $\mathscr{A}$-index defined in the previous chapter.

But in reality, it is too hypothetical to assume that a DM is absolutely pessimistic or absolutely optimistic. On the contrary, it is quite reasonable to assume that a DM is pessimistic when he is composed of more of pessimism and less of optimism. Similarly, a DM, who is optimistic may be reasonably assumed to be composed of more of optimistic and less of pessimistic attitude and performs accordingly in a decision vacuum. Again, it can be seen that in a particular decision vacuum two pessimistic (or optimistic) DMs differ due to the variant level of pessimism-biasness naturally belong to them.

So, in order to develop a pessimistic (or, optimistic) DM's choice pattern, we consider level of pessimism as a key parameter in this analysis. Level of pessimism can be represented well in linguistic terms and hence, may be defined as a linguistic variable. We define the set of 'terms' on it as

$$\{ \cdots, \text{very very low}, \ \text{very low}, \ \text{low}, \ \text{moderate}, \ \text{high}, \ \text{very high}, \cdots \}.$$

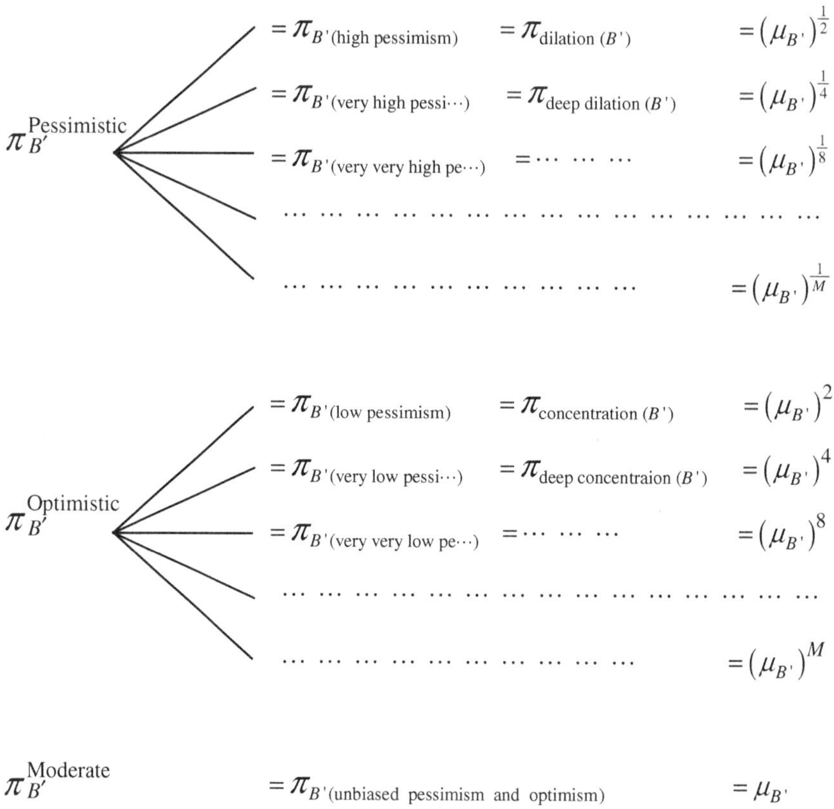

Fig. 4.2. A fuzzy semantic rule using the most frequently used modifiers – concentration and dilation.

In the fuzzy literatures (Zimmermann (1991)) subjectively valued different terms are often defined by modifier (or hedge) which is an operation that modifies the meaning of a linguistic term or more generally, of a fuzzy set. Therefore, for a sequence of values of $p = [(1/M), M]$, a semantic rule can be generated in order to define linguistic modifier (or hedge) for fuzzy set $\pi_{B'}$ under various degrees of pessimism. Here we use the most frequently used modifiers – concentration and dilation, as is given in Fig. 4.2.

Now, in order to avoid complexity, from the above semantic rule, we take out only the following cases: For a pair of intervals (X, B) in S_2, the grade of rejection of B is defined (Fig. 4.3),

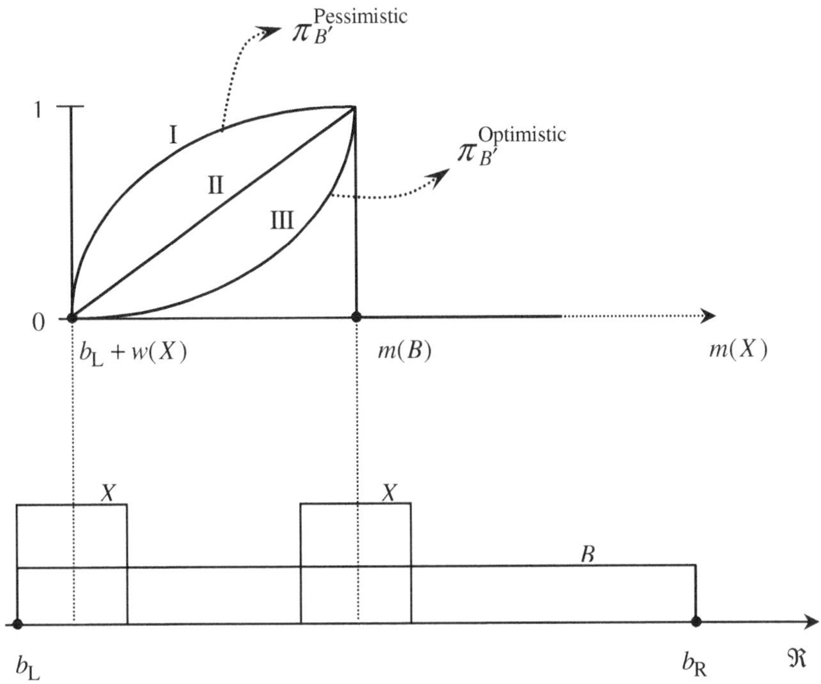

Fig. 4.3. Membership functions of *rejection of B* for the pair of intervals (*X*, *B*) in S_2, where, (I) is for a pessimistic, (II) is for a moderate and (III) is for an optimistic DM.

either,

as a pessimistic DM's membership-function, $\pi_{B'}^{\text{Pessimistic}} = (\mu_{B'})^{1/2}$,

or,

as an optimistic DM's membership-function, $\pi_{B'}^{\text{Optimistic}} = (\mu_{B'})^2$,

or,

as a moderate DM's membership-function, $\pi_{B'}^{\text{Moderate}} = \mu_{B'}$,

and consequently, the grade of rejection of *X* is defined by

$$\pi_{X'} = 1 - \pi_{B'}.$$

Therefore, for a particular DM with a specific degree of pessimism, the maximal non-dominated pair (X^0, B) in S_2, is defined as (see Figs. 4.4 and 4.5),

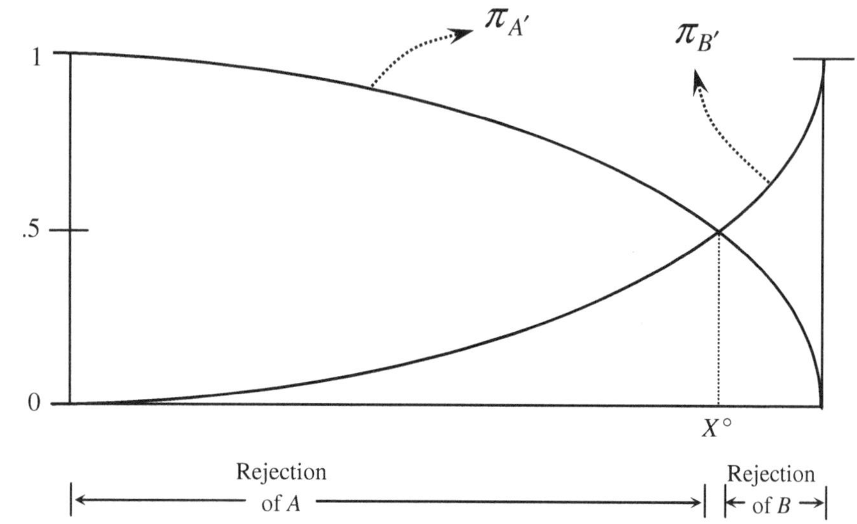

Fig. 4.4. An Optimistic DM's preference pattern
(used the same axes as of Figs. 4.1 and 4.3).

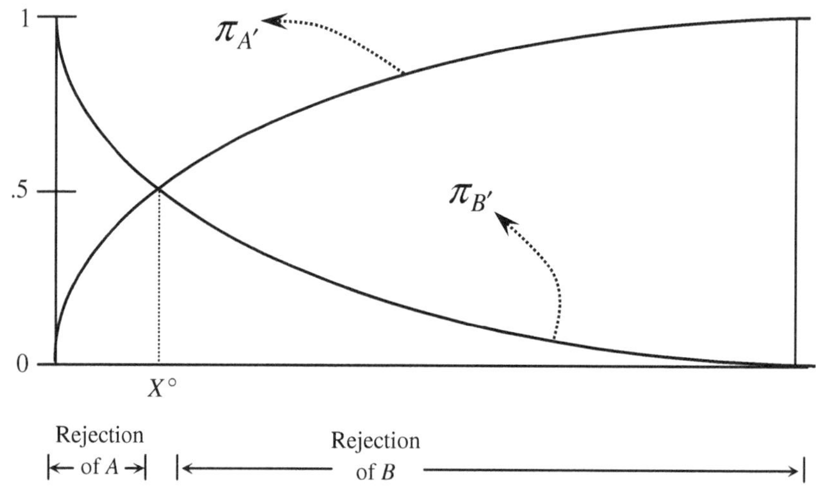

Fig. 4.5. A Pessimistic DM's preference pattern
(used the same axes as of Figs. 4.1 and 4.3).

$$\pi_{\text{nond}}(X^0)_{\substack{X^0 \in X \\ (X,B) \in S_2}} = \sup_{(X,B)\in S_2} \pi_{\text{nond}}(X)$$

$$= \sup_{\substack{X^0 \in X \\ (X,B) \in S_2}} \min\{\pi_{X'}(X),\ \pi_{B'}(X)\}.$$

For a pair of intervals (X, B) in S_2, where B is given, for a particular DM the condition of indifference (non-dominance) is achieved when the degree of *rejection of B* is *approximately equal* to degree of *rejection of X*, or, symbolically, when,

$$\pi_{B'} \cong \pi_{X'},$$

or, when,

$$\pi_{B'} \cong 1 - \pi_{B'},$$

or, when,

$$\pi_{B'} \cong 0.5.$$

The notation ' $\cong$' denotes fuzzy equivalence and linguistically may be interpreted as *is approximately equal to* or *is close to* or *is close enough to.*

Similarly, for any given pair (X_1, B), by the extent of deviation of the value $\pi_{B'}(X_1)$ from the pivotal value 0.5, the degree of rejection of B (or, X_1) or, acceptability of X_1 (or, B) can be explained at least in linguistic expressions which is the most comfortable as well as the most practical way of expressing an opinion.

A linguistic scale may be defined as is given in Fig. 4.6 and on the basis of which linguistic or subjective assessment in favour of a decision may be made.

4.4.1. *Illustrative example*

Let us refer again Example 3.2.1.4. Here we consider A and B_3 are two alternative profit intervals where

$$A = [120,\ 180] = \langle 150,\ 30 \rangle$$

and

$$B_3 = [110,\ 230] = \langle 170,\ 60 \rangle.$$

We have to find out the most preferred alternative from a pessimistic or the so-called risk-averse DM's point of view.

As we argued earlier that pessimism or risk averseness in reality happens to be in variety. In a risky situation it is not very hard to find two pessimistic (or two optimistic) DMs who decide differently. Therefore, by argument, a certain *intensity of risk preference* or a certain *degree of pessimism* (or *optimism*) turns out to be a responsible factor there that guides the DM to find his own choice. And in the line of this argument, we have called a DM as *pessimistic* who is inclined more towards pessimism than towards optimism (or in other words, a pessimistic DM has comparatively low intensity of risk preference). Similarly, we have called a DM as *optimistic* one who has less pessimism and more optimism (or has high intensity of risk preference).

Here, we compute the degree of rejection of B_3 from among A and B_3 from a pessimistic DM's point of view (assuming $p = \frac{1}{2}$):

$$\pi_{B_3'}^{\text{Pessimistic}} = \pi_{B_3'} \text{ (high pessimism)}$$

$$= \left(\frac{m(A) - ((b_3)_L + w(A))}{m(B_3) - ((b_3)_L + w(A))} \right)^{\frac{1}{2}}$$

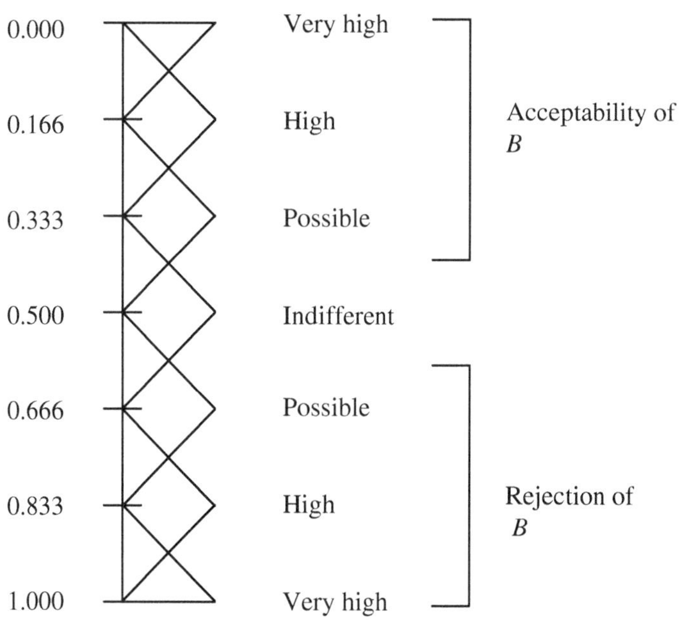

Fig. 4.6. A linguistic scale

$$= \left(\frac{150-(110+30)}{170-(110+30)} \right)^{\frac{1}{2}}$$

$$= \left(\frac{10}{30} \right)^{\frac{1}{2}} = 0.58.$$

Using the numerical scale given in Fig. 4.6, this result implies that the DM prefers to remain indifferent to rejection and acceptance of B_3.

Now, we compute the degree of rejection of B_3 from among A and B_3 from an optimistic DM's point of view (assuming $p = 2$):

$$\pi_{B_3'}^{\text{Optimistic}} = \pi_{B_3'} \text{ (low pessimism)}$$

$$= \left(\frac{10}{30} \right)^{2} = 0.11.$$

Using the numerical scale given in Fig. 4.6, the result implies that this DM highly prefers acceptance of B_3.

If we wish to compute rejection of B_3 for a pessimistic DM with very high degree of pessimism, then we take $p = \frac{1}{4}$ and

$$\pi_{B_3'}^{\text{Pessimistic}} = \pi_{B_3'} \text{ (very high pessimism)}$$

$$= \left(\frac{10}{30} \right)^{\frac{1}{4}} = 0.76.$$

This implies that this DM prefers high rejection of B_3 and hence high acceptance of A.

Similarly, for an optimistic DM with very low degree of pessimism, we take $p = 4$ and hence

$$\pi_{B_3'}^{\text{Optimistic}} = \pi_{B_3'} \text{ (very low pessimism)}$$

$$= \left(\frac{10}{30} \right)^{4} = 0.01,$$

which implies that this DM prefers B_3 with a very high acceptability.

4.5. Preference pattern for a minimization problem

In order to develop a specific preference pattern between a pair of profit-intervals, a fuzzy preference ordering scheme for both pessimistic and optimistic DMs is discussed in the preceeding sections of this chapter. Now, we assume that A and B are two interval-valued costs and the problem here is to choose the minimum cost-interval from among them.

When we talk about a choice out of an inexact (imprecise) and uncertain environment, there is definitely some room for discussion on a DM's personal preference and on his/her risk-taking ability which happens to be in his/her decision-making potential. As a whole, the main question to be answered is: Can we interpret the term *minimum cost* as the *best cost* from a DM's personal preference point of view – whatever be the DM – a risk-averse or a risk-lover, a pessimistic or an optimistic one. In another words, we need to clarify our goal: we are going to find out *a satisfactory minimum cost* (*to a particular DM*) from among the available alternatives.

In the preceeding two sections we have developed a *fuzzy preference ordering* for a *maximization problem*. Here in this section we briefly review the same preference ordering for a *minimization problem environment* and the result is further used in some methodologies developed in the following chapters.

4.5.1. Assumptions for the minimization problem:

1. Low cost is better than high cost.
2. More certainty is better than less certainty.
3. If less cost is associated with more uncertainty, a DM makes a trade-off between the two.
4. To a pessimistic DM, assumption 2 is somewhat more important than assumption 1. To an optimistic DM, assumption 1 is somewhat more important than assumption 2.

Definition 4.5.2. Any pair of intervals A and B are ordered as (A, B) if $m(A) \le m(B)$ and are classified into two sets, $\mathcal{T}_1$ and $\mathcal{T}_2$ as follows:

(i) $(A, B) \in \mathcal{T}_1$, if $w(A) \le w(B)$

(ii) $(A, B) \in \mathcal{T}_2$, if $w(A) > w(B)$

Then, for a *minimization problem*,

- For $(A,B) \in \mathcal{T}_1$, unless A and B are identical, *A is always the best choice.*
- For $(A,B) \in \mathcal{T}_2$, *fuzzy preferences between A and B* may be constructed.

Definition 4.5.3. For a given interval, $A = [a_L, a_R] = < m(A),\ w(A) >$ there exists a set of intervals (A,Y) in $\mathcal{T}_2$, where, $Y = [y_L, y_R] = < m(Y),\ w(Y) >$ is a variable interval. Hence, we define a fuzzy set A' as the *rejection of A* for any pair (A,Y) in $\mathcal{T}_2$, where, $A' = \{(A,\ Y) \,/\, m(A) \le m(Y),\ w(A) > w(Y)\}$ with the membership function $\mu_{A'}(A,Y)$, $\mu_{A'}$ being a function $\mathcal{T}_2 \to [0,1]$, as follows:

$$\mu_{A'}(A,Y) = \begin{cases} 1 & \text{if } m(Y) = m(A), \\ \max\left\{0, \dfrac{(a_R - w(Y)) - m(Y)}{(a_R - w(Y)) - m(A)}\right\} & \text{if } m(A) \le m(Y) \le a_R - w(Y), \\ 0 & \text{otherwise.} \end{cases}$$

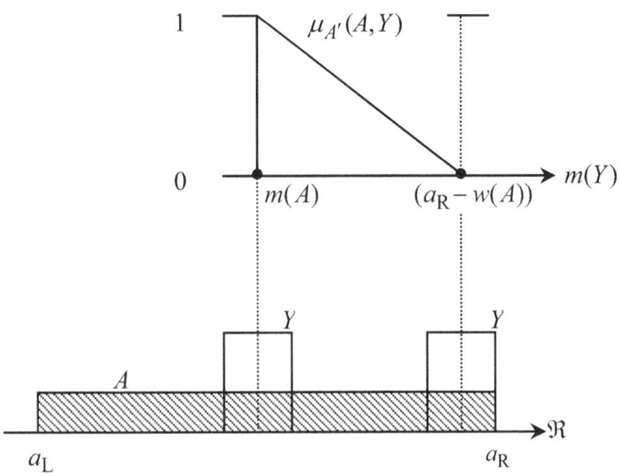

Fig. 4.7. The Membership function of *rejection of A* for the pair of intervals (A,Y) in $\mathcal{T}_2$.

We can interpret different values of $\mu_{A'}$ as follows:

if $\mu_{A'} = 1$, then A is definitely rejected,

if $\mu_{A'} = 0$, then A is definitely accepted and

if $\mu_{A'} \in \,]0,1[$ (i.e., $\mu_{A'}$ takes a value between 0 and 1 excluding 0 and 1), then this defines different grades of *rejection of* A (Fig. 4.7).

Now, for consistency and completeness in the formulation of preference structure for any pair (A,Y) in $\mathcal{T}_2$, let us define another fuzzy set, Y', the *rejection of* Y in $\mathcal{T}_2$ as *the complement of* A', i.e.,

$$\forall Y \text{ of } (A,Y) \in \mathcal{T}_2, \qquad \mu_{Y'}(A,Y) = 1 - \mu_{A'}(A,Y)$$

Then, for any set of intervals (A,Y) in $\mathcal{T}_2$, the ordered pair of fuzzy sets (A',Y') is defined as a *fuzzy partition* between their preferences (Lowen (1978)). In other words, in order to find out *satisfactory minimum* (Zimmermann (1991), Yen & Langari (1999)) from among (A,Y) in $\mathcal{T}_2$, $\mu_{A'}$ describes the degree of dominance of A over Y and $\mu_{Y'}$ describes the degree of dominance of Y over A.

Thus, there exists a fuzzy set of non-dominated preference, *nond* for (A,Y) in $\mathcal{T}_2$, such that

$$nond = A' \cap Y',$$

where,

$$\mu_{\substack{nond \\ (A,Y) \in \mathcal{T}_2}} = \min\{\mu_{A'}(A,Y),\ \mu_{Y'}(A,Y)\}$$

A DM hence may choose the pair of intervals (A,Y) in $\mathcal{T}_2$ as the pairs of non-dominated preferences if μ_{nond} is close enough to

$$\sup_{(A,Y) \in \mathcal{T}_2} \mu_{nond}.$$

Now, let us define a *fuzzy preference ordering for different DMs with different degrees of pessimism.* The degrees of pessimism may be interpreted as the intensities of risk preferences, i.e., a high degree of pessimism may be interpreted as a low intensity of risk preference and a low degree of pessimism as a high intensity of risk preference. We modify the construction of A'.

Definition 4.5.4. Let us define A' as *the rejection of A for a DM with a specific degree of pessimism* with the non-linear membership function (see Section 4.4),

$$\pi_{A'} = \left(\mu_{A'}\right)^p, \quad \mu_{A'} \in [0,1]$$

where, the parameter p is the (non-fuzzy) level of pessimism for any DM and $p \in [M^{-1}, M]$, where M is a positive finite large number.

When the value of $p = M^{-1}$, the DM is said to be an absolutely pessimistic person who practically does not take any risk. Similarly, for an absolutely optimistic DM, to get the membership function for rejection of A, we use $p = M$. If we use $p = 1$, we get back to the definition 4.5.3. This level of pessimism, i.e., at $p = 1$, may be regarded as the reflection of a moderate DM's attitude against the aforementioned extreme odds.

In reality, it is too hypothetical to assume that a DM is absolutely pessimistic or absolutely optimistic. On the contrary, it is quite reasonable to assume that a DM is said to be pessimistic when he exhibits more pessimism and less optimism. Similarly, a DM, who is optimistic, may be reasonably assumed to have more of an optimistic and less of a pessimistic attitude. Again, it can be seen that in a particular decision vacuum two pessimistic (or optimistic) DMs may differ due to their natural levels of pessimism-biasness. Therefore, for a sequence of values of $p \in [M^{-1}, M]$, a semantic rule can be generated (a detail discussion is given in Section 4.4) in order to define linguistic modifier (or hedge) for fuzzy set $\pi_{A'}$ under various degrees of pessimism.

4.6. Comparative advantage of the Fuzzy Preference Ordering over the other interval ordering schemes

The preference ordering between any two intervals can be classified in the following way.

When two intervals A and B are

1) non-overlapping (disjoint), i.e., $a_R < b_L$, then there exists a strict preference relation between A and B.

2) partially overlapping, i.e., $a_L < b_L \le a_R < b_R$, there exists a strict preference relation between A and B.

3) such that B is nested in A, i.e., $B \subseteq A$, then

a) if $m(A) = m(B)$ and $w(A) \geq w(B)$, then there exists a strict preference relation between A and B. For obtaining the minimum (as well as the maximum) from among A and B, (unless $B \equiv A$) B is always preferred to A.

b) if $m(A) \leq m(B)$ and $w(A) \geq w(B)$, then there exists a strict preference relation between A and B. To be specific, *for obtaining a maximum alternative between the two*, B is strictly preferred to A.

c) if $m(A) \geq m(B)$ and $w(A) \geq w(B)$, then also there exists a strict preference relation between A and B. *For obtaining a minimum alternative between the two*, B is strictly preferred to A.

d) for finding a minimum from case 3(b) and finding a maximum from case 3(c), *soft* or *fuzzy ordering* is defined between the alternatives.

An interval has four characteristic parameters, viz., left limit, right limit, mean and width. However, an interval can be denoted by any two and only two of these parameters. For example, Ishibuchi & Tanaka (1990)'s $\leq_{LR}$ relation uses only left and right limits but $\leq_{mw}$ relation uses only mean and width of the comparable intervals. Okada & Gen (1994) and $\mathscr{A}$-index use only mean and width but the *fuzzy preference ordering* scheme uses all of the four parameters in the definition of its membership function. In fact, we have used all of the four parameters even in any single piece of analysis for the sake of better representation and understanding.

Next, we need to clarify the basis (or principle) of the above classification. While choosing any two intervals A and B, we have kept an interval fixed and moved the other interval freely from right to left. The variable interval has undergone a change both in its position and width with respect to the fixed interval in order to curb the unnecessary and obvious repetition (as we see in some of the past as well as recent literatures (Nakahara et al (1992), Sun & Yao (2008)). Above classification facilitates the discussion on interval ranking by simultaneous yet practical consideration of cardinal (value-based) and ordinal (preference-based) ranking approaches. This classification also aids explanation and comparison of the existing indices of the literature.

Two transitive order relations given by Moore (1979) cannot explain the ranking between overlapping intervals (i.e., paragraphs 2 and 3 of the above classification). Ishibuchi & Tanaka (1990)'s ranking methodology, which is considerably advanced over those of Moore (1979), can be valid for all of the paragraphs stated above except the para 3(d) of the above classification.

Kundu (1997)'s *fuzzy leftness relationship* for comparing any two intervals on the real line can solve the Ishibuchi & Tanaka's problem but it is deficient in another way. For any two intervals A and B, if $m(A) = m(B)$ and $w(A) \neq w(B)$, which corresponds to para 3(a) of the above classicification, Kundu's approach indicates that both A and B are optimal (non-dominated) choices which is clearly inconsistent to our intuition.

Okada & Gen (1994) extended Ishibuchi & Tanaka's methodology and defined an interface between partial and total orderings on the basis of α, β's different values. They intended to reduce the number of incomparable pairs of intervals (in para 3(d)). However, the logical basis for choosing between partial and total orderings was not discussed. In addition when total ordering is opted by Okada & Gen (1994) for resolving para 3(d), neither the uniqueness nor the effectiveness of such comparison was discussed. Also no light has been shown on whether the ordering as discussed is acceptable to any DM, say, a risk taker or a risk averse DM.

Two Index-Comparing Tables – *Index Comparing Table I* (pages 78-81) and *Index Comparing Table II* (page 88) are given in this section. *Index Comparing Table I* is a huge table broken into four pages. Table 4.1a (left panel) and Table 4.1a (right panel) are simltaneously horizontally linked; and Table 4.1b (left panel) and Table 4.1b (right panel) are vertical continuations of previous two panels.

In the *Index Comparing Table I*, the cases for which Ishibuchi & Tanaka (1990)'s approach gives incomparable pairs are also not clearly comparable by Okada & Gen (1994)'s approach because of too many options on the α and β's choice and particularly due to the absence of clear guideline about the delimiting level between partial and total orderings.

Whereas, our *acceptability index* ($\mathscr{A}$*–index*) for comparing the values of any two interval numbers can be applied to any of the cases stated above. $\mathscr{A}$–index further defines how much higher one interval is than another in terms of the DM's level of objective satisfaction. However, the main weakness of the $\mathscr{A}$–index is that it is only a value-based ranking index and it can be only partially applicable from a pessimistic (or a risk averse) DM's point of view. So, for improving the applicability and competency of the index, even as a pessimistic DM's preference index, a *fuzzy preference relation* for a class of intervals is introduced.

The *Index Comparing Table I* shows the advantage of the fuzzy preference ordering over the other approaches in terms of its flexibility, completeness and clarity in obtaining a decision. Here the term 'flexibility' takes into account a DM's risk-intensity, the term 'completeness' signifies comparability between any pair of intervals and the term 'clarity' indicates a straightforward conclusion in an inexact decision situation.

Index Comparing Table II (page 88) describes a similar table in a much concised format but its intention is to compare the results of different indices for five different cases for both maximization and minimization problem situation. Here we also include some recent but relevant indices which have appeared in the literature referring the existing indices discussed above and Chapters 2 and 3.

	Interval *B* glides, but, not *A*	**Graphical Representation**	**Ishibuchi & Tanaka (1990)**	**Okada & Gen (1994)**
1	$A = [10, 20] = <15, 5>$ $B = [22, 24] = <23, 1>$	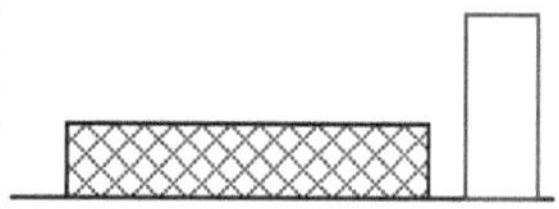	$A \le_{LR} B$ $\Rightarrow B$ is preferred.	$A \preceq_{0,1} B$ $\Rightarrow B$ is preferred.
2	$A = [10, 20] = <15, 5>$ $B = [20, 22] = <21, 1>$	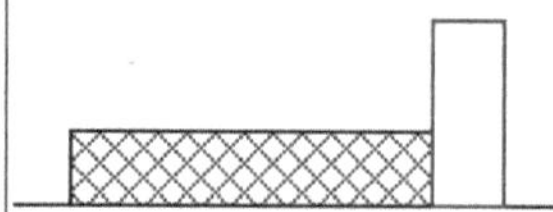	$A \le_{LR} B$ $\Rightarrow B$ is preferred.	$A \preceq_{0,1} B$ $\Rightarrow B$ is preferred.
3	$A = [10, 20] = <15, 5>$ $B = [19, 21] = <20, 1>$	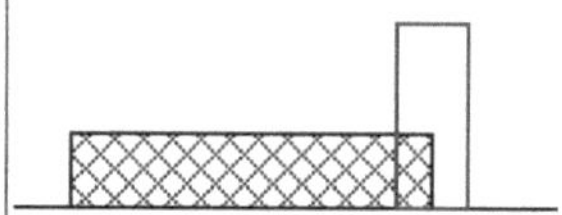	$A \le_{LR} B$ $\Rightarrow B$ is preferred.	$A \preceq_{0,1} B$ $\Rightarrow B$ is preferred.
4	$A = [10, 20] = <15, 5>$ $B = [18, 20] = <19, 1>$	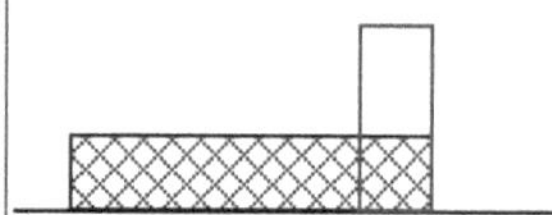	$A \le_{LR} B$ $\Rightarrow B$ is preferred.	$A \preceq_{0,1} B$ $\Rightarrow B$ is preferred.
5	$A = [10, 20] = <15, 5>$ $B = [17, 19] = <18, 1>$	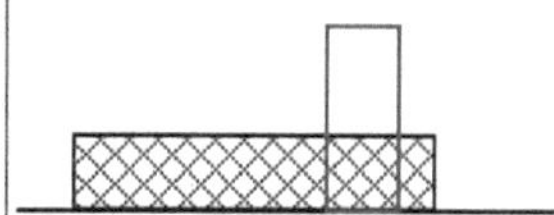	$A \le_{mw} B$ $\Rightarrow B$ is preferred.	$A \preceq_{0,1} B$ $\Rightarrow B$ is preferred.

Table 4.1a (left panel): Index Comparing Table I

	Kundu (1997)	**$\mathscr{A}$ –index**	**Fuzzy Preference ordering**
1	$P_{AB}(x < y) = 1$ Left $(A,B) = 1$ $\Rightarrow A$ is left to B.	$\mathscr{A}(A \olessthan B) > 1$ $\Rightarrow$ B is superior.	Impersonally *B* is preferred to *A*
2	$P_{AB}(x < y) = 1$ Left $(A,B) = 1$ $\Rightarrow A$ is left to B.	$\mathscr{A}(A \olessthan B) = 1$ $\Rightarrow B$ is superior.	Impersonally *B* is preferred to *A*
3	$P_{AB}(x < y) =.98$ Left $(A,B) =.95$ $\Rightarrow A$ is left to B.	$\mathscr{A}(A \olessthan B) = .83$ $\Rightarrow$ B is superior.	Impersonally *B* is preferred to *A*
4	$P_{AB}(x < y) =.90$ Left $(A,B) =.80$ $\Rightarrow A$ is left to B.	$\mathscr{A}(A \olessthan B) = .66$ $\Rightarrow$ B is superior.	Impersonally *B* is preferred to *A*
5	$P_{AB}(x < y) =.80$ Left $(A,B) =.60$ $\Rightarrow A$ is left to B.	$\mathscr{A}(A \olessthan B) = .5$ $\Rightarrow$ B is superior.	Impersonally *B* is preferred to *A*

Table 4.1a (right panel): Index Comparing Table I

	Interval *B* glides, but, not *A*	**Graphical Representation**	**Ishibuchi & Tanaka (1990)**	**Okada & Gen (1994)**
6	$A = [10, 20] = <15, 5>$ $B = [14, 16] = <15, 1>$	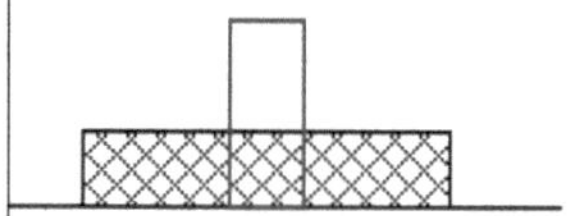	$A \le_{mw} B$ $\Rightarrow$ *B* is preferred	$A \preceq_{0,1} B$ $\Rightarrow$ *B* is preferred
7	$A = [10, 20] = <15, 5>$ $B = [13, 15] = <14, 1>$	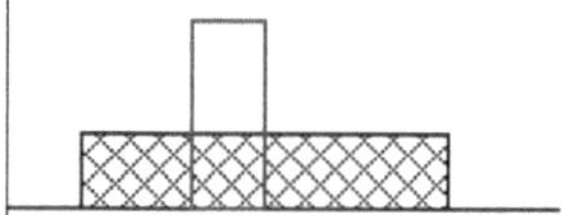	Incomparable situations	$A \preceq_{\alpha,\beta} B$ for $\alpha = .25$, $\beta = .75$
8	$A = [10, 20] = <15, 5>$ $B = [12, 14] = <13, 1>$	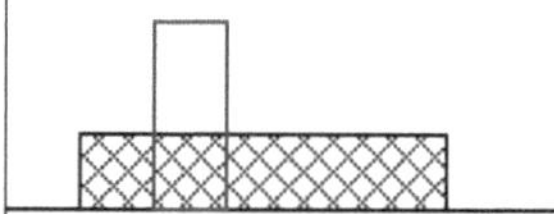		$A \preceq_{\alpha,\beta} B$ for $\alpha = .5$, $\beta = .5$
9	$A = [10, 20] = <15, 5>$ $B = [11, 13] = <12, 1>$	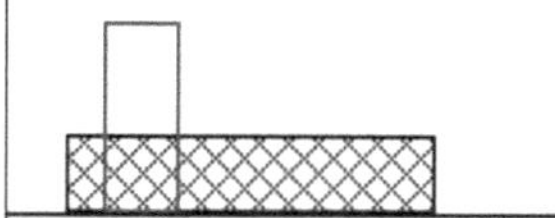		$B \preceq_{\alpha,\beta} A$ for $\alpha = .75$, $\beta = .25$
10	$A = [10, 20] = <15, 5>$ $B = [10, 12] = <11, 1>$		$B \le_{LR} A$ $\Rightarrow$ *A* is preferred.	$B \preceq_{0,1} A$ $\Rightarrow$ *A* is preferred.

Table 4.1b (left panel): Index Comparing Table I

	Kundu (1997)	$\mathscr{A}$ –index	Fuzzy Preference ordering		
			For a Pessimistic DM, $p = ½$	**For a rational DM, $p = 1$**	**For an Optimistic DM, $p = 2$**
6	$P_{AB}(x<y)=.5$ Left $(A,B)=0$ ⇒ Both A and B are optimal choice.	$\mathscr{A}(A \ominus B)=0$ ⇒ B is not superior to A but B is preferable, ($\because w(A) > w(B)$)	Impersonally B is preferred to A		
7	$P_{AB}(x<y)=.4$ Left $(A,B)=0$ Left $(B,A)=.2$ ⇒ B is left to A.	$\mathscr{A}(B \ominus A)=.16$ ⇒ A is superior.	Acceptability of B is high	Acceptability of B is fairly high.	Indifference between A and B
8	$P_{AB}(x<y)=.3$ Left $(A,B)=0$ Left $(B,A)=.4$ ⇒ B is left to A.	$\mathscr{A}(B \ominus A)=.33$ ⇒ A is superior.	Acceptability of B is fairly high	Indifference between A and B	Acceptability of A is high
9	$P_{AB}(x<y)=.2$ Left $(A,B)=0$ Left $(B,A)=.6$ ⇒ B is left to A.	$\mathscr{A}(B \ominus A)=.5$ ⇒ A is superior.	Indifference between A and B	Acceptability of A is fairly high.	Acceptability of A is very high
10	$P_{AB}(x<y)=.1$ Left $(A,B)=0$ Left $(B,A)=.8$ ⇒ B is left to A.	$\mathscr{A}(B \ominus A)=.66$ ⇒ A is superior.	Impersonally A is preferred to B		

Table 4.1b (right panel): Index Comparing Table I

4.6.1. *A note on some recent ranking schemes*

Here we discuss three relevant works published after 2000-01.

Let $A = [a_L, a_R]$ and $B = [b_L, b_R]$ be any two *interval utilities*. Wang et al (2005) defined the degree of one interval utility being greater than another one as the degree of preference. Here, the preference of B over A is symbolically denoted as $P(A \prec B)$ and is defined as

$$P(A \prec B) = \frac{\max(0, b_R - a_L) - \max(0, b_L - a_R)}{(b_R - b_L) + (a_R - a_L)}.$$

Similarly, the degree of preference of A over B is defined as

$$P(B \prec A) = \frac{\max(0, a_R - b_L) - \max(0, a_L - b_R)}{(a_R - a_L) + (b_R - b_L)}$$

Therefore, for any given A and B,

$$P(A \prec B) + P(B \prec A) = 1.$$

It is further defined that

if $P(A \prec B) > .5 + \delta$, then B is superior to A to the degree of $P(A \prec B)$,

if $P(A \prec B) < .5 - \delta$, then B is inferior to A to the degree of $P(A \prec B)$ and

if $P(A \prec B) \in [.5 - \delta, .5 + \delta]$, then B is said to be indifferent to A

where, δ is a very small positive number predetermined by the DM.

This index uses only left and right limit in its definition. Its output is almost similar to that of Kundu (1997)'s *fuzzy leftness relationship* (which is based on probabilistic statement of intervals) except a slight deviation on a particular situation where the degree of preference is at the vicinity of .5 leaving just at .5. On the contrary, $\mathscr{A}$-index, which is defined with mid-point and width of the intervals, gives very much likely result to this index.

$P(A \prec B)$ and $\mathscr{A}(A \circledcirc B)$ are linearly dependent if both are mapped onto same domain.

Wang et al (2005)'s $P(A \prec B)$ index assumes intervals A and B as interval utilities. So, we can compare $P(A \prec B)$ with our *fuzzy preference ordering* scheme for the maximization problem and kundu (1997)'s *fuzzy leftness relationship* on an identical platform. See Figs. 4.8, 4.9 and 4.10. For each of the three figures, here we have taken a wide interval A as a reference interval and interval B, a much narrower interval than A, as a variable interval with variable a mean and a constant width k, as follows:

$$A = [a_L, a_R] = \langle m(A), w(A) \rangle \text{ and}$$

$$B = [b_L, b_R] = \langle m(B), k \rangle, \text{ such that } k \ll w(A).$$

On the horizontal axis, three parameters of interval A, respectively the left limit a_L, the mean value $m(A)$ and right limit a_R, are spotted. Now, variable interval B starts moving from $b_R \geq a_L$ towards the right up to the point $a_R \leq b_L$. In Fig. 4.8, the vertical axis measures *acceptability of A* and *acceptability of B* in two opposite directions. It is to be noted that a particular direction from 0 to 1 on the vertical axis should be read as either acceptability of an interval or rejectability of the other.

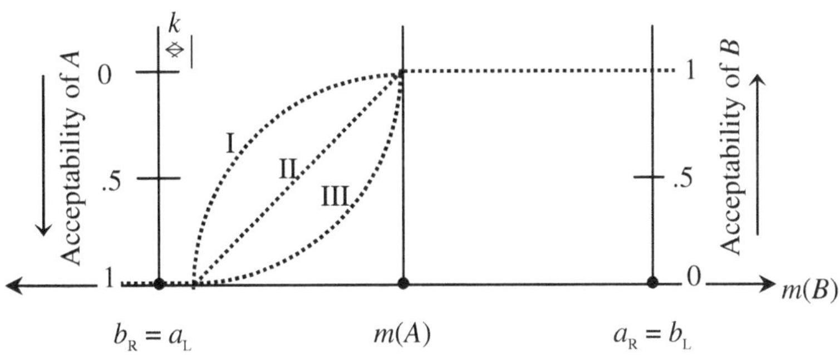

Fig. 4.8. Pessimistic, moderate and optimistic DM's preference patterns (curves I, II and III respectively) for utility intervals according to the fuzzy preference ordering scheme (Sengupta & Pal's 2nd index).

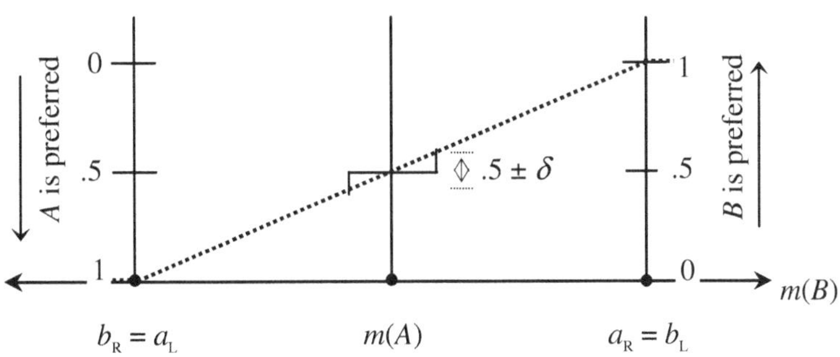

Fig. 4.9. Preference pattern for utility intervals according to Wang et al (2005).

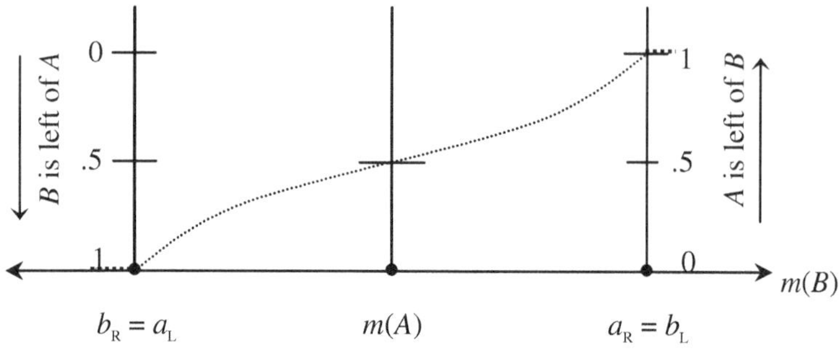

Fig. 4.10. Preference pattern for utility intervals according to Kundu (1997).

In Fig. 4.8, three curves, may be called the *acceptability curves*, are drawn with bold dots. All of the three curves have three parts each. Initial and final parts are horizontal lines which are common. Different middle parts signify fuzzy membership functions for different risk attitude of the DM (as was discussed in Section 4.4). *Acceptability curve I* interprets pessimistic DM's level of preference for *B*, *curve II* interprets a moderate DM's acceptability and *curve III* interprets an optimistic DM's choice for *B*. Here, the level of preference, level of acceptability and choice level are used as synonymous terms.

Fig. 4.9 shows Wang et al (2005)'s $P(A \prec B)$ with the diagonal dotted line. Horizontal axis is defined as above. Here the vertical axis denotes preferability of an interval over the other.

Fig. 4.10 depicts Kundu (1997)'s $P_{AB}(x < y)\big| x \in A, y \in B$ where the vertical axis measures leftness of an interval with respect to another. Here also the horizontal axis is same as above.

Jiang et al (2008) used *possibility degree of interval numbers* (PDIN) for the treatment of uncertain inequality constraints. PDIN represents a possibility degree with which one interval is larger or smaller than another. Jiang et al (2008) here extended the *probability method to compare interval numbers* and proposed the construction of possibility degree $P_{A \le B}$ and $P_{A \ge B}$ based on three relations of interval numbers *A* and *B* of the following types:

(a) $a_L \le a_R \le b_L \le b_R$

(b) $a_L \le b_L \le a_R \le b_R$

(c) $b_L \le a_L \le a_R \le b_R$

We find these three sets of relations among the Nakahara et al (1992)'s six cases described in Section 2.3.2 (Figs. 2.1 (*i*), (*ii*) & (*iv*)). Nakahara et al (1992) defined the probability value for all the six cases in a *generalized format.*

However, Jiang et al (2008)'s PDIN combines above three individual cases into one as follows:

$$P_{A\le B}=\begin{cases}1, & b_L \ge a_R\\ \dfrac{b_R-a_R}{b_R-b_L}+\dfrac{a_R-b_L}{b_R-b_L}\cdot\dfrac{b_L-a_L}{a_R-a_L}+\frac{1}{2}\cdot\dfrac{a_R-b_L}{b_R-b_L}\cdot\dfrac{a_R-b_L}{a_R-a_L}, & a_L \le b_L < a_R \le b_R\\ \dfrac{b_R-a_R}{b_R-b_L}+\frac{1}{2}\cdot\dfrac{a_R-a_L}{b_R-b_L}, & b_L < a_L < a_R \le b_R.\end{cases}$$

See also Figs. 2.2 ((*i*), (*ii*) & (*iv*)).

As $P_{A\le B}$ + $P_{A\ge B}$ = 1, we have,

$$P_{A\ge B}=\begin{cases}0, & b_L \ge a_R\\ \frac{1}{2}\cdot\dfrac{a_R-b_L}{b_R-b_L}\cdot\dfrac{a_R-b_L}{a_R-a_L}, & a_L \le b_L < a_R \le b_R\\ \dfrac{b_L-a_L}{a_R-a_L}+\frac{1}{2}\cdot\dfrac{a_R-a_L}{b_R-b_L}, & b_L < a_L < a_R \le b_R\end{cases}$$

As compared to Kundu (1997)'s fuzzy leftness (or rightness) relationship between any two interval numbers, Jiang et al (2008)'s PDIN does not contradict the former in result but gives an explicit and easier format to work with.

In a very recent paper Sun & Yao (2008) has given a set-theoretic approach of ranking two interval numbers which appears as following format (if we maintain our notational standard):

$$P([A]\le[B])=\min\left\{\max\left(\frac{b_R-a_L}{2\left(w(B)+w(A)\right)},0\right),1\right\}$$

$$P([B]\le[A])=\min\left\{\max\left(\frac{a_R-b_L}{2\left(w(A)+w(B)\right)},0\right),1\right\}$$

Obviously $P([A]\le[B])\in[0,1]$ and $P([A]\le[B])+P([B]\le[A])=1$.

This index is entitled as *possibility degree function* (PDF) for comparing two intervals and is almost similar to Wang et al (2005)'s index in terms of both definition and result. However, Wang et al (2005)'s work is remarkable only in the cases where it compares equi-centered or closely-centered intervals and declare them 'indifferent intervals' in terms of preference. Kundu (1997) concluded that only the equi-centered intervals are of 'indifferent' or 'optimal' or 'non-dominated' choice. Whereas fuzzy preference ordering scheme gives a complete interval ranking model and defines different sets of 'pairs of intervals' for which there exist *strict* and *fuzzy preference relation* and *indifference* between the interval-attributes. If finding a preference pattern is the target, it must be defined clearly whose preference pattern is to be sketched, whose purpose is to be served, whether he is a rational human being or a pessimistic or optimistic investor. Fuzzy preference ordering scheme qualifies to address all of these queries.

	Assumptions of Minimization problem	**Assumptions of Maximization problem**
1	*Low cost* is better than *high cost.*	More money is better than less money.
2	More certainty is better than less certainty.	
3	If *less cost* is associated with more uncertainty, a DM makes a trade-off between the two.	If *more money* is associated with more uncertainty, a DM undergoes a trade-off between the two.
4	To an optimistic DM, assumption 1 is somewhat more important than assumption 2. To a pessimistic DM, assumption 2 is somewhat more important than assumption 1.	

Table 4.2: Comparison of assumptions for the minimization and maximization models of fuzzy preference ordering

Another noteworthy point is of considering maximizing and minimizing goal while a preference judgment is to be made amongst interval attributes. Ishibuchi & Tanaka (1990)'s outstanding work first addressed this realistic and intuitive feature by using $\leq_{mw}$ and $\leq^{*}_{mw}$ relations. Most of the researchers in interval preference ranking, as we find, have not given proper attention to this issue. $\mathscr{A}$-index considers this issue (proposition 3.2.7) with proper stress. The fuzzy preference ordering scheme uses this proposition as one of its base lines.

Another recent work (Majumdar & Bhunia (2007)) may be quoted in this regard which reveals interval ranking by tallying only the left limits or only the centers of the intervals concerned. The process may be considered simple and easy to implement but remembering Freeling (1980)'s warning,

> "...... by reducing the whole of our analysis to a single number, we are losing much of the information we have purposely been keeping throughout our calculations",

we are on the view that in an uncertain environment, the preference relation for two interval attributes should be defined through a fuzzy membership function. An interval ranking method should be able to represent preference relations for interval numbers in linguistic or fuzzy terms (some fuzzy literatures (Tong & Bonissone (1980), Yuan (1991) etc.) may be referred in this regard). For example, instead of simply concluding that A is preferred or not preferred to B, the method should be able to represent the degree of preference, such as A highly dominates B, A is slightly preferred to B, or A and B are more or less the same, etc.

The *Index Comparing Table II* (Table 4.3 in page 88) frames the indices in brief but a simultaneous reference of maximization and minimization goal is incorporated.

But, before that, in order to facilitate clear understanding, an assumption tally for both the goals is referred in Table 4.2.

Problem 4.6.1.1. Let for interval $A = [2,12] = \langle 7,5 \rangle$, there exist a set of intervals, $B = \{B_1, B_2, B_3, B_4, B_5\}$ such that

$B_1 = [12,14] = \langle 13, 1 \rangle$,

$B_2 = [9,11] = \langle 10, 1 \rangle$,

$B_3 = [6,8] = \langle 7, 1 \rangle$,

$B_4 = [3,5] = \langle 4, 1 \rangle$ and

$B_5 = [0,2] = \langle 1, 1 \rangle$.

Let's try to compare each of the Bs with respect to interval A such that Case 1 deals with A and B_1; Case 2 deals with A and B_2; and so on. Please note that Case 4 here deals with the comparison between A and B_4, which is referred earlier by Majumdar & Bhunia (2007).

In the middle most column of Table 4.3, bold-lettered words correspond to cases defined in problem 4.6.1.1. and the words in italics correspond to the paragraphs denoting classes in the classification of intervals (defined at the beginning of Section 4.6). In this column, an arrow relates a classification clause with a particular example under minimization or maximization objective.

For Min Problems			For Max Problems	
According to Sengupta & Pal's 2nd index	**According to Ishibuchi & Tanaka, Okada & Gen, Kundu, Sengupta & Pal's 1st index**		**According to Ishibuchi & Tanaka, Okada & Gen, Kundu, Sengupta & Pal's 1st index**	**According to Sengupta & Pal's 2nd index**
A is superior to B_1.	A is superior to B_1.	**Case 1** *Para 2*	B_1 is superior to A.	B_1 is superior to A.
For *different DMs*, *Fuzzy Preferences* are defined between A and B_2	**Different authors have different opinions; some of them coincide**	**Case 2** *Para 3(c)* → ← *Para 3(d)*	B_2 is superior to A.	B_2 is superior to A.
B_3 is superior to A	B_3 is superior to A (Kundu: both A and B_3 are optimal choice)	**Case 3** *Para 3(a)*	B_3 is superior to A (Kundu: both A and B_3 are optimal choice)	B_3 is superior to A
B_4 is superior to A.	B_4 is superior to A.	**Case 4** *Para 3(c)* → ← *Para 3(d)*	**Different authors have different opinions; some of them coincide**	For *different DMs*, *Fuzzy Preferences* are defined between A and B_4
B_5 is superior to A.	B_5 is superior to A.	**Case 5** *Para 2*	A is superior to B_5.	A is superior to B_5.

Table 4.3: Index Comparing Table II.

4.7. References

R.E. **Bellman** & L.A. **Zadeh (1970)**, Decision-making in a fuzzy environment, *Management Science* 17 (4): 141 – 164.

A.N.S. **Freeling (1980)**, Fuzzy sets and decision analysis, *IEEE Trans. Systems Man Cybernet.* 6: 341-354.

H. **Ishibuchi** & H. **Tanaka (1990)**, Multiobjective programming in optimization of the interval objective function, *European Journal of Operational Research* 48 (2): 219 –225.

C. **Jiang** , X. Han, G.R. Liu & G.P. Liu (**2008**), A nonlinear interval number programming method for uncertain optimization problems, *European Journal of Operational Research* 188: 1–13.

S. **Kundu (1997)**, Min-transitivity of fuzzy leftness relationship and its application to decision making, *Fuzzy Sets and Systems* 86 (3): 357–367.

R. **Lowen (1978)**, On fuzzy completeness, *Information Science* 14: 107–113.

J. **Majumdar** & A.K. **Bhunia (2007)**, Elitist genetic algorithm for assignment problem with imprecise goal, *European Journal of Operational Research*177 (2): 684-692.

R.E. **Moore (1966)**, *Interval Analysis*, Prentice-Hall, Englewood Cliffs, New Jersey.

R.E. **Moore (1979)**, *Method and Application of Interval Analysis*, SIAM, Philadelphia.

Y. **Nakahara,** M. Sasaki & M. Gen (**1992**), On the linear programming problems with interval coefficients, *Computers and Industrial Engineering* 23(1-4): 301-304.

S. **Okada** & M. **Gen (1994)**, Order Relation between Intervals and Its Application to Shortest Path Problem, *Japanese journal of Fuzzy Theory and Systems* 6 (6): 703 – 717. (In English).

H.L. **Sun** & W.X. **Yao (2008),** The basic properties of some typical systems' reliability in interval form, *Structural Safety* 30: 364–373.

R.M. **Tong** and P.P. **Bonissone (1980)**, A linguistic approach to decision making with fuzzy sets, *IEEE Trans. Systems Man Cybernet.* 10: 716-723.

Y.M. **Wang**, J.B. Yang & D.L. Xu (**2005**), A preference aggregation method through the estimation of utility intervals *Computers & Operations Research* 32: 2027–2049.

J. **Yen** & R. **Langari (1999)**, *Fuzzy Logic – Intelligence, Control and Information*, Pearson Education, Inc.

Y. **Yuan (1991)**, Criteria for evaluating fuzzy ranking methods, *Fuzzy Sets and Systems* 44: 139-157.

H.J. **Zimmermann (1991)**, *Fuzzy Set Theory and its Application*, Kluwer Academic Publishers, Boston/Dordrecht/London.

Chapter 5

Solving the Shortest Path Problem with Interval Arcs

5.1. Introduction

This chapter presents an algorithm for the shortest path problem with the connected arcs in a transportation network and its objective is to figure out the fastest route to travel from one node to another node. Each arc in the network represents transportation time or cost in traveling and, as a consequence, may fluctuate considerably depending on traffic condition, payload etc. In this problem, we define each arc in terms of real intervals, which can be regarded as the simplest variation of an imprecise or a fuzzy value and can better be defined here as a composed pair of optimistic and pessimistic values (which are left and right limits respectively) (Okada & Gen (1994)) or as a composed pair of Best- and Worst-Case of economic consequences (Schjaer-Jacobsen (1996, 2002)).

Though the interval arithmetic has attracted many a researchers, the practical problem of ranking of interval numbers, the application of interval ranking and the applicability of interval numbers as a whole in decision problems, have not received the attention that they merit. Here, Dijkstra's algorithm has been redesigned to handle the case in which most of the parameters are inexact and given in terms of simple intervals. Okada & Gen (1994) proposed an algorithm to solve the

A. Sengupta and T.K. Pal: Fuzzy Preference Ordering, STUDFUZZ 238, pp. 91–109.
springerlink.com

shortest path problem, which tends to generate a set of incomparable intervals as non-dominated solutions of the problem.

Okada & Gen (1994) extended Ishibuchi & Tanaka (1990)'s methodology and defined an interface between partial and total orderings. Therefore, the incomparable intervals so generated were as same as the outcome of Ishibuchi & Tanaka's ranking approach. Okada & Gen (1994) first tried to resolve the problem of incomparability of intervals using a ranking strategy of their own but their method seems to work in an ad hoc manner and so the result is not always unique and self-explanatory when used in an algorithm for solving the shortest path problem. In Chapter 4, we proposed our 2nd interval ranking index, *fuzzy preference ordering of intervals* that gives a clearer and more convincing decision than that of Okada & Gen (1994). There we have discussed *fuzzy preference ordering* for the *maximization* and *minimization problem setup* which are actually the logical reverse cases of one another. Here in this chapter, we find an application of the ranking methodology in a shortest path problem with interval arcs.

Section 5.2 gives a procedure for choosing a preferred minimum from a set of n intervals and Section 5.2.1 gives an illustration of the procedure with a suitable numerical example. A shortest path problem with connected arcs in interval data is defined and the extended Dijkstra's algorithm is presented and discussed in Section 5.3. Next in Section 5.4, in order to compare our results with those obtained by Okada & Gen (1994), we consider the same problem as in their work. The final section contains some concluding remarks.

5.2. Choosing a preferred minimum from a set of intervals

In the previous chapter we defined and explained fuzzy preference ordering for two interval-attributes in a minimization framework. On the basis of that discussion here is given a step-by-step procedure (Sengupta & Pal (2006)) for selecting a preferred minimum from a set of n intervals, or a best preferred interval from a set of n cost intervals.

Step 1: Rewrite the given set of intervals U_i, $\forall i \in \{1, 2, \dots n\}$ both in [left limit, right limit] and $\langle$mean, width$\rangle$ form.

Step 2: As an initial step, order them according to their ascending mid-values. Next, if the intervals are equi-centered, i.e.,

$$m(U_i) = m(U_j),\ \forall i, j \in \{1, 2, \dots n\},$$

then order U_i and U_j according to ascending $w(U_i)$ and $w(U_j)$.

(It is to be noted that in this step, intervals can also be ordered according to descending mid-values. In that case subsequent changes have to be made in the following steps.)

Step 3: Rename *ordered intervals* as $X_1, X_2, \ldots X_n$.

Step 4: Form a table of order $n \times n$, where each cell entry x_{ij} represents a preference relation between the ordered pair of intervals (X_i, X_j). This table is called a *Fuzzy Rejection Table* for the minimization problem (Table 5.1).

Step 5: If Step 2 initially orders the intervals according to ascending mid-values, then cells (i, j), $\forall i \leq j$, of the *fuzzy rejection table* are classified in $\mathcal{T}_1$ and $\mathcal{T}_2$ according to *definition 4.5.2* and the ordered pairs of intervals (X_i, X_j) corresponding to these cells are *directly* comparable. On the contrary, the interval pairs corresponding to the rest of the cells are *not directly* comparable (refer *definitions 4.5.2* and *4.5.3*).

	X_1	X_2	......	X_j	...	X_n
X_1	$\pi_{X_1'}(X_1, X_1)$	$\pi_{X_1'}(X_1, X_2)$	......	$\pi_{X_1'}(X_1, X_j)$		$\pi_{X_1'}(X_1, X_n)$
X_2	$\pi_{X_2'}(X_2, X_1)$	$\pi_{X_2'}(X_2, X_2)$	......	$\pi_{X_2'}(X_2, X_j)$		......
...	......	......	......	......	...	......
X_i	$\pi_{X_i'}(X_i, X_1)$	Grade of rejection of X_i over X_2	......	Grade of rejection of X_i over X_j		......
...			......			......
X_n	$\pi_{X_n'}(X_n, X_1)$	Grade of rejection of X_n over X_2	......	Grade of rejection of X_n over X_j	...	$\pi_{X_n'}(X_n, X_n)$

Table 5.1: Interpretation of the content of the cells in the *Fuzzy Rejection table* for an Interval Minimization Problem.

Step 6(a): In cell (i, j), $\forall i = j$, we need to compare the rejectability (or acceptability) of an interval, X_i with itself. Logically this case is a special case of comparison of set $\mathcal{T}_2$ (refer *definition 4.5.2*). Here we set $\pi_{X_i'}(X_i, X_i)=0.5$, as these intervals are not differentiable in terms of rejection or acceptability.

Step 6(b): Cell (i, j), $\forall i < j$, defines preferability between the ordered pair of intervals (X_i, X_j) as follows (also refer *definitions 4.5.3 & 4.5.4*):

□ If $(x_i)_R \leq (x_j)_L$,

then set $\pi_{X_i'}(X_i, X_j)=0$

(here, $\pi_{X_i'}(X_i, X_j)$ denotes the membership *grade of rejection of X_i over X_j*),

□ If $m(X_i) = m(X_j)$ and

- if $w(X_i) = w(X_j)$,

  then assign $\pi_{X_i'} = .5$,

- if $w(X_i) < w(X_j)$,

  then assign $\pi_{X_i'} = 0$,

- if $w(X_i) > w(X_j)$,

  then assign $\pi_{X_i'} = 1$

□ If $m(X_i) < m(X_j)$ and

- if $w(X_i) \leq w(X_j)$,

  then assign $\pi_{X_i'} = 0$,

- if $w(X_i) > w(X_j)$, then compute

$$\pi_{X_i'} = \max\left\{0, \left(\frac{(x_i)_R - w(x_j) - m(x_j)}{(x_i)_R - w(x_j) - m(x_i)}\right)^p\right\}.$$

Step 6(c): Cell (i, j), for any specific $i > j$, is a complementary cell to the cell (j,i).

So, the complementary membership value of cell (j,i) should be assigned to cell (i, j), i.e.,

$$\pi_{X_i'}(X_i, X_j) = 1 - \pi_{X_j'}(X_j, X_i).$$

Step 7: After completion of computation for all the cells (i, j), $i, j \in \{1, 2, \ldots n\}$, the *fuzzy rejection table for the minimization problem* will look like Table 5.3. Now, each cell of the table will contain a membership value signifying the *grade of rejection of the row element compared to the column element.* Consequently, each row of the table will contain an array of membership values indicating the *grade of rejection of the row element compared to the corresponding column elements.*

Step 8: Compute the final stages as:

I. *Iteration 1*: Add the membership values row-wise. We get n results for the n rows. Place the results in $(n+1)^{\text{th}}$ column of the table.

II. Select the row with minimum row-sum and the row heading, for example, $X_i^* = X_k \in \{X_1, X_2, \ldots X_n\}$, containing the minimum row-sum is the preferred minimum from among $X_1, X_2, \ldots X_n$.

III. *Iteration 2*: Strike the corresponding row as well as column out from the table and repeat addition of the membership values row-wise to select the next preferred minimum from among the remaining $X_1, X_2, X_3 \ldots$.

IV. *Iteration 3*: Repeat the process until all the intervals are ordered according to their preferences to the evaluator.

Step 9: Decode X_i^* in terms of U_i's and the process ends here.

5.2.1. *Numerical illustration of the procedure*

Let us consider the following set of cost intervals:

$$A = [4,\ 18] = \langle 11,\ 7\rangle$$

$$B = [7,\ 20] = \langle 13.5,\ 6.5\rangle$$

$$C = [13,\ 17] = \langle 15,\ 2\rangle$$

$$D = [10,\ 18] = \langle 14,\ 4\rangle$$

$$E = [6,\ 17] = \langle 11.5,\ 5.5\rangle$$

$$F = [10,\ 16] = \langle 13,\ 3\rangle.$$

To order the given set of intervals according to the decreasing order of preferences (for finding minimum from among the set), the above procedure is illustrated as follows:

5.2.1.1. First, the intervals are ordered according to their ascending mean value level and renamed as:

A	E	F	B	D	C		
X_1	X_2	X_3	X_4	X_5	X_6	⟵	Renamed

5.2.1.2. Now, *Fuzzy Rejection Table* of 6×6 cells is constructed (refer Table 5.2). The cells on the principal diagonal alignment which correspond to pairs of identical elements and the cells to the right of the principal diagonal are to be classified first into two sets $\mathcal{T}_1$ and $\mathcal{T}_2$.

> ***Observation***: As stated above, Step 2 of the procedure defines an initial ordering of the intervals in ascending order of their mid-values. Cells to the right of the principal diagonal contain those ordered pair of intervals, which can be classified according to *definition 4.5.2*. On the contrary, pair of intervals in the cells to the left of the principal diagonal cannot be classified directly according to this definition.
>
> Let us take an example of two complementary cells, say *cell (3,5)* and *cell (5,3)*, which are respectively from the right and left of the principal diagonal of the *Fuzzy Rejection Table*. *Cell (3,5)* corresponds an ordered pair of intervals (X_3, X_5) and *cell (5,3)* corresponds to (X_5, X_3). Therefore, in this case, the pair of intervals (X_5, X_3) of *cell (5,3)* is not directly comparable according to our *definition 4.5.2*, but we can compare the same pair of intervals of *cell (3,5)* where the intervals are ordered as (X_3, X_5). (If initially the intervals were ordered in their descending mid-values, then the cells on the principal diagonal and the cells to the left of the principal diagonal would have been classified into $\mathcal{T}_1$ and $\mathcal{T}_2$. Refer *definition 4.5.2*). Therefore,

Classify *Cell (i, j)*, $\forall\ i = j$ and $i < j$, into two sets $\mathcal{T}_1$ and $\mathcal{T}_2$ and mark accordingly. The rest of the cells are left unmarked (Table 5.2).

5.2.1.3. In the next step, the membership values are assigned to cells (Step 6 of Section 5.2). All the cells on the principal diagonal are assigned 0.5 and the remaining cells, marked as $\mathcal{T}_1$, are given 0. Membership values are calculated (considering $p = 1$) for $\mathcal{T}_2$ cells. Now, the *Fuzzy Rejection Table* looks like a *right triangular matrix* (refer Table 5.2).

	X_1	X_2	X_3	X_4	X_5	X_6
X_1	$\mathcal{T}_1$ .5	$\mathcal{T}_2$ .67	$\mathcal{T}_2$ .5	$\mathcal{T}_2$ 0	$\mathcal{T}_2$ 0	$\mathcal{T}_2$ .2
X_2		$\mathcal{T}_1$ .5	$\mathcal{T}_2$ .4	$\mathcal{T}_1$ 0	$\mathcal{T}_2$ 0	$\mathcal{T}_2$ 0
X_3			$\mathcal{T}_1$ .5	$\mathcal{T}_1$ 0	$\mathcal{T}_1$ 0	$\mathcal{T}_2$ 0
X_4				$\mathcal{T}_1$ .5	$\mathcal{T}_2$ .8	$\mathcal{T}_2$ .67
X_5					$\mathcal{T}_1$ .5	$\mathcal{T}_2$ .5
X_6						$\mathcal{T}_1$ .5

Table 5.2: The initial *Fuzzy Rejection Table* for the example given in Section 5.2.1.

	X_1	X_2	X_3	X_4	X_5	X_6
X_1	.5	.67	.5	0	0	.2
X_2	.33	.5	.4	0	0	0
X_3	.5	.6	.5	0	0	0
X_4	1	1	1	.5	.8	.67
X_5	1	1	1	.2	.5	.5
X_6	.8	1	1	.33	.5	.5

Table 5.3: The complete *Fuzzy Rejection Table*, where cell (i, j) contains a fuzzy membership value to define rejectability of the ith interval compared to the jth one.

	X_1	X_2	X_3	X_4	X_5	X_6	Iteration 1	Iteration 2	Iteration 3	Iteration 4	Iteration 5	
X_1	.5	.67	.5	0	0	.2	*1.87*	*1.20*	*.70*			
X_2	.33	.5	.4	0	0	0	*1.23*					
X_3	.5	.6	.5	0	0	0	*1.60*	*1.00*				
X_4	1	1	1	.5	.8	.67	*4.97*	*3.97*	*2.97*	*1.97*	*1.17*	
X_5	1	1	1	.2	.5	.5	*4.20*	*3.20*	*2.20*	*1.20*		
X_6	.8	1	1	.33	.5	.5	*4.13*	*3.13*	*2.13*	*1.33*	*.83*	
							X_2	X_3	X_1	X_5	X_6	X_4

Table 5.4: The Final Selection Table for the problem given in Section 5.2.1.

5.2.1.4. The remaining complementary cells are assigned complementary values. That is, if any cell (*i, j*) contains $\alpha_{ij} \in [0,1],\ \forall i, j \in \{1,2,...n\}$, then, the complementary cell (*j, i*) will have the value $\alpha_{ij} = (1-\alpha_{ij})$. Thus, the complete *Fuzzy Rejection Table* is formed (Table 5.3).

5.2.1.5. Now, the selection operation is done as is described in Step 8 of Section 5.2 and Table 5.4. X_2 comes out here as the best minimum from the six alternatives. Consequently, in absence of interval X_2 in the set, the next best minimum is X_3 and so on. Ultimately, here we get the order of intervals from the best (satisfactory) minimum to the worst (unsatisfactory) minimum (considering $p = 1$) as $X_2, X_3, X_1, X_5, X_6, X_4$.

5.2.1.6. If a *pessimistic* DM's choice is to be framed at $p = ½$, then from similar computation we get the result $X_3, X_2, X_1, X_6, X_5, X_4$ and if a *very pessimistic* DM's choice is to be framed i.e., at $p = ¼$, then the result becomes $X_3, X_2, X_6, X_1, X_5, X_4$. Similarly, different optimistic DM's results can also be obtained at values of p = 2, 4, and so on.

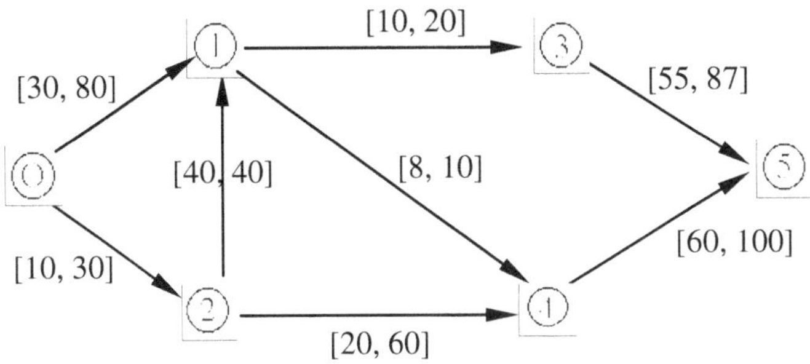

Fig. 5.1. The network problem with interval arcs

5.3. Shortest Path Problem

Although the shortest path problem is one of the simple network problems, it has a wide range of applications. The objective of the problem is to find the most economical path, say, with minimum total time or cost from an origin to the destination.

As the time and cost fluctuate depending on traffic conditions, payload etc., imprecise or uncertain data of some kind are to be utilized as values of arcs. However, it is not always easy for a DM to specify the membership function or probability distribution in an inexact environment. At least in some of the cases, the use of interval coefficients may serve the purpose better. Here, we consider such an example (Fig. 5.1) earlier used by Okada & Gen (1994).

To solve the problem, we modify Dijkstra's algorithm to incorporate interval parameters as data. To present the modified algorithm, we need the following notation:

p : an optimism/pessimism parameter (defines intensity of risk preference)

P : set of nodes with permanent labels

T : set of nodes with temporary labels

U_j : set of minimum length(s) expressed in intervals from node 0 to node j

d_{ij} : distance between nodes i and j.

Iteration 1

Set $p = 1$

$U_0 = \{[0, 0]\}$,

$U_1 = \{[30, 80]\}$, $U_2 = \{[10, 30]\}$

$U_3 = U_4 = U_5 = \{+\infty\}$,

$P = \{0\}$

$T = \{1, 2, 3, 4, 5\}$

$\min\{U_1, U_2\} = \min\{d_{01}, d_{02}\}$

$= \min\{[30, 80], [10, 30]\}$

$= [10, 30] \subset U_2{}^*$

(node 2 becomes the permanent node.)

Iteration 2

$P = \{0, 2\}$

$U_2 = \{[10, 30]\}$

$T = \{1, 3, 4, 5\}$

$U_1 = \min\{d_{01}, U_2 + d_{21}\}$

$= \min\{[30, 80], [10, 30] + [40, 40]\}$

$= \min\{[30, 80], [50, 70]\}$

$= [50, 70]$

(Here the rule of interval minimization at p = 1 has been applied.)

$U_4 = U_2 + d_{24} = \{[10, 30] + [20, 60]\} = \{[30, 90]\}$

$U_3 = U_5 = \{+\infty\}$

$\min\{U_1, U_4\} = \{[50, 70]\} \subset U_1{}^*$

(Node 1 becomes the next permanent node.)

Iteration 3

$P = \{0, 2, 1\}$

$U_1 = \{[50, 70]\}$

$T = \{3, 4, 5\}$

$U_3 = \{U_1 + d_{13}\} = \{[50, 70] + [10, 20]\} = \{[60, 90]\}$

$U_4 = \min\{U_2 + d_{24}, U_1 + d_{14}\}$

$= \min\{[30, 90], [50, 70] + [8, 10]\}$

$= \min\{[30, 90], [58, 80]\}$

$= [58, 80]$

(Here the rule of interval minimization has been applied.)

$U_5 = \{+\infty\}$

$\min\{U_3, U_4\} = \{[60, 90], [58, 80]\} = U_4{}^*$

(Node 4 becomes the next permanent node.)

Iteration 4

$P = \{0, 2, 1, 4\}$
$U_4 = \{[58, 80]\}$
$T = \{3, 5\}$
$U_3 = \{U_4 + d_{43}\}$: non-existent.
$U_5 = \{U_4 + d_{45}\} = \{[58, 80] + [60, 100]\} = \{[118, 180]\}$

- Return to the previous iteration, i.e., *Iteration 3*.
- Ignore the minimization operation and
- Retain both the paths $\{U_3, U_4\}$ as optimum and *consider both as permanent nodes* and repeat *Iteration 4*.

Iteration 4.1 (*Iteration 4* repeated)

$P = \{0, 2, 1, (3, 4)\}$
$U_3 = \{[60, 90]\}$
$U_4 = \{[58, 80]\}$
$T = \{5\}$
$U_5 = \min\{U_3 + d_{35}, U_4 + d_{45}\}$
$= \min\{[60, 90] + [55, 87], [58, 80] + [60,100]\}$
$= \min\{[115,177], [118,180]\}$
$= [115,177] = \{U_3 + d_{35}\}^*$
(*Node 5 through d_{35} becomes the next and final permanent node.*)

Iteration 5

$P = \{0, 2, 1, (3), 5\}$
$T = \{\varnothing\}$
The process ends.

Therefore, taking $p = 1$ (i.e., assuming that the DM is a moderate one), the Shortest Path is obtained: $0 \rightarrow 2 \rightarrow 1 \rightarrow 3 \rightarrow 5$.

A pictorial description of the solution of the problem given in Fig. 5.1 is shown in Fig. 5.2. In Fig. 5.2, the progress towards solution at each node is given *in the box.* Here, a "✱" denotes local optimality at a particular stage and a "~~strike through~~" defines an elimination of a path found uneconomical after some progress.

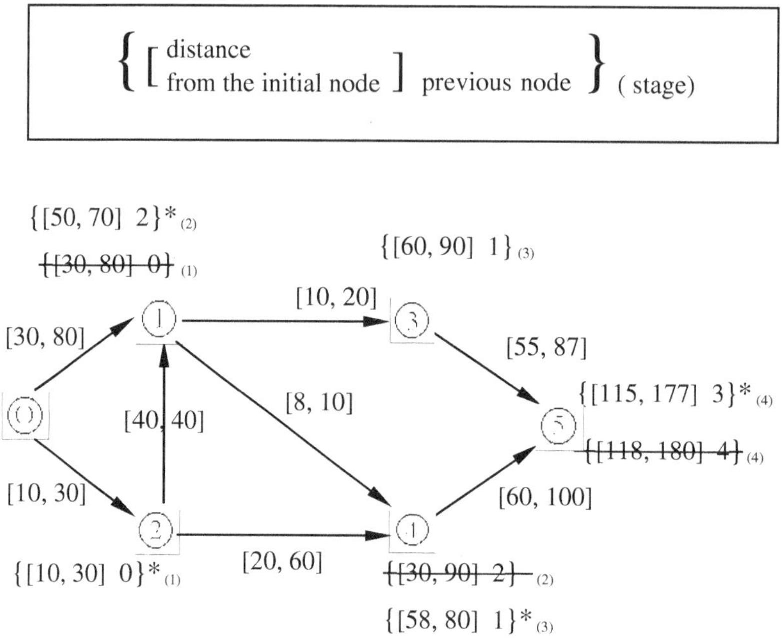

Fig. 5.2. The progressive solution of the problem given in Fig. 5.1.

In contrast the approach of Okada & Gen (1994) initially gave three non-dominated solutions, *viz.*,

Route I.	$0 \to 2 \to 4 \to 5$	total route length = [90, 190]
Route II.	$0 \to 1 \to 3 \to 5$	total route length = [95, 187]
Route III.	$0 \to 2 \to 1 \to 3 \to 5$	total route length = [115, 177].

The values of their algorithm-parameters α, β were reset by trial-and-error and the steps were repeated again and again. Finally, at $\alpha = .7$ and $\beta = .3$, because of incomparability, route II and III are obtained as two satisfactory minimal length paths and presented to the DM. Moreover, Okada & Gen did not rule out further possibility of choosing from among the existing routes and so, if smallest right bound is to be given priority then route III is selected and if smallest center of the intervals is to be given priority then route I is selected. Okada & Gen gave much importance to the selection of the values of the parameters α and β between 0 and 1, whereas the role of a particular DM's attitude (say for example, a DM is *very very optimistic* or the DM may be *slightly pessimistic*) and its impact on the selection of the values of α and β were not discussed.

Firstly we do not agree with the statement that route II and III are incomparable (though in course of our solution procedure we have not encountered the need for comparing these routes). According to our proposed definition of fuzzy preference ordering (*definition 4.5.4*) for selection of best minimum from among the route II & III, we may have the following results.

a) For an *optimistic* DM (at $p = 2$), route II & III are regarded as equally good.

b) For a rational DM (at $p = 1$), route III is possibly selected.

c) For a *pessimistic* DM (at $p = ½$), route III is highly selected.

Secondly, unlike Okada & Gen (1994), the result we have obtained here is distinctively clear. In our process, the DM is not presented a set of non-dominated solutions to select from. Here we have shown that every pair of intervals is comparable. A DM may be indifferent in choosing between two or more solutions but this 'indifference' is entirely due to his rational behaviour. As has been discussed, depending on whether a DM is more pessimistic or more optimistic, a fuzzy region of indifference is determined on the basis of his perception and intensity of bias (inclination) toward risk. For any particular comparison, two or more solutions are either equivalent or can be ordered strictly or weakly according to the DM's personal attitude. For any certain predefined bias of the DM, the preference order of intervals may change but a satisfactorily distinct solution is bound to come.

For example, if the solution is sought for an *optimistic* DM (at $p = 2$), then we have the optimized route as $0 \to 2 \to 4 \to 5$ and the total route length is [90, 190]. Even for a *very optimistic* DM (at $p = 4$), we have the same route. However, for a *pessimistic* DM (at $p = ½$), we have the optimized route as $0 \to 2 \to 1 \to 3 \to 5$ where the total route length is [115, 177].

Thirdly, we need not resort to trial-and-error in any part of our procedure. So, our algorithm can solve the network problem with fur-fewer iterations than that of Okada & Gen (1994).

5.4. Large-Scale numerical example

In order to further compare our results with those obtained by Okada & Gen (1994), we now consider the large problem given here. The problem was to obtain a path to travel from Boston (No.1) to Los Angeles (No.70) in the shortest time by car according to the map given in Fig. 5.3. The road network consists of 70 cities and 211 connecting links between cities and their associated driving times. The driving times between cities connected by lines were expressed as intervals with a width equal to about 20% of the average driving time (unit: minute). However, in the paper (Okada & Gen (1994)), driving times between the cities other than those on the nine obtained paths were omitted. We accept the data (as is given in the cited reference) of nine route lengths which were obtained according to their method by setting $\alpha = 1$ and $\beta = 0$. [Basically at $\alpha = 1$ and $\beta = 0$, the interval ranking methodology of Okada & Gen (1994) is reduced to Ishibuchi & Tanaka (1990)'s ranking methodology. For interval ranking we do not question the validity of Ishibuchi & Tanaka (1990)'s finding. However our main query lies beyond this juncture. As we have stated in Section 4.6 the question is whether the point given in 3(d) can be explained by these methodologies. In this sense Okada & Gen (1994) deserve credit for their method that attempted to address para 3(d) of classification of interval numbers (Section 4.6) towards the building of complete interval ranking methodology].

Fig. 5.3. US map of driving times between major cities (Okada & Gen (1994))

In this light we initiate the problem of finding the shortest route from among the obtained nine routes (Okada & Gen (1994)). Table 5.5 (given in pages 106-107) contains the interval data for the nine routes. The intervals denoting the routes are all nested intervals (refer point 3 of Section 4.6) and are so arranged that the means be in ascending order but the widths be in descending order.

We apply our method given in Section 5.2 for finding a preferred minimum from a set of 9 nested intervals. For a rational DM we obtain the best (minimum) route as A_3 from the nine alternatives and the worst (minimum) route as A_9 with the complete ordering,

$$A_3 < A_4 < A_2 < A_5 < A_6 < A_7 < A_1 < A_8 < A_9.$$

Consequently, in absence of A_3, here the next best minimum is A_4 and so on. The entire tabular work for final selection is given in Table 5.6 (page 108). This final selection table shows a moderate DM's preferences at $p = 1$.

We also note that if the selection is made for a pessimistic DM (at $p = ½$), we obtain the best route as A_4 and the worst route as A_1 with the following ordering

$$A_4 < A_3 < A_7 < A_6 < A_2 < A_5 < A_8 < A_9 < A_1$$

and if the selection is made for an optimistic DM (at $p = 2$), we have the best route as A_2 and worst route as A_9 in the following order

$$A_2 < A_3 < A_4 < A_1 < A_5 < A_6 < A_7 < A_8 < A_9.$$

We also note that Okada & Gen (1994), by setting α and β closer to each other, i.e., $\alpha = .7$ and $\beta = .3$, arrive at 3 optimum routes, viz., A_2, A_3 and A_4. Again they suggest another two options such as A_9 (where shortest right limit becomes the order of intervals) and A_1 (where smallest center becomes the order of intervals) by setting the values of (α, β) as (1, 1) and (0, 0) respectively. Here, the DM is still presented with many options among which he must choose, whereas our solution gives clear decision considering the DM's personal preference.

5.5. Conclusion

This chapter defines a solution procedure for a shortest path problem in which the network arcs are represented as interval numbers. Dijkstra's algorithm has been generalized to incorporate interval parameters as data. Consequently, the main issue dealt with was developing a suitable ranking strategy or preference structure for a set of interval numbers. Impersonal as well as prejudiced fuzzy preference orderings between two interval numbers were defined earlier in Chapter 4. This work has been extended for ranking a set of intervals simultaneously and executed in a network problem. With the ranking strategy developed here, Dijkstra's algorithm has been generalized and made simple and effective for finding permanent nodes from among the temporary nodes at every stages of iteration. Though imprecision is embedded in the problem definition, the solution procedure is well capable of providing an impersonal solution as well as many prejudiced solutions on the basis of subjective interaction only (and not by any objective query) with the human decision maker.

Path	Length of path	Path and length of each connecting line										
A_1	[2826, 4430] = <3628, 802>	1	→	3	→	23	→	33	→	44	→	
			[205, 313]		[645, 1105]		[232, 343]		[245, 367]		[63, 119]	
		45	→	58	→	70						
			[686, 1028]		[750, 1150]							
A_2	[2860, 4406] = <3633, 773>	1	→	3	→	23	→	33	→	46	→	
			[205, 313]		[645, 1105]		[232, 343]		[502, 702]		[249, 373]	
		56	→	58	→	70						
			[277, 415]		[750, 1150]							
A_3	[2924, 4384] = <3654, 730>	1	→	8	→	9	→	12	→	23	→	
			[155, 257]		[273, 435]		[183, 275]		[313, 419]		[232, 348]	
		33	→	46	→	56	→	58	→	70		
			[502, 702]		[249, 373]		[277, 415]		[750, 1150]			
A_4	[2938, 4380] = <3659, 721>	1	→	8	→	9	→	12	→	40	→	
			[155, 257]		[273, 435]		[183, 275]		[691, 937]		[130, 194]	
		41	→	55	→	63	→	70				
			[517, 775]		[722, 1082]		[277, 415]					
A_5	[3053, 4377] = <3715, 662>	1	→	8	→	9	→	12	→	23	→	
			[155, 257]		[273, 435]		[183, 275]		[313, 419]		[232, 348]	
		33	→	46	→	56	→	58	→	62	→	70
			[502, 702]		[249, 373]		[277, 415]		[485, 627]		[394, 516]	

Table continued to next page -

Path	Length of path	Path and length of each connecting line										
A_6	[3080, 4372] = <3726, 646>	1	→	3	→	12	→	23	→	33	→	
			[205, 313]		[552, 652]		[313, 419]		[232, 348]		[502, 702]	
		46	→	56	→	58	→	70				
			[249, 373]		[277, 415]		[750, 1150]					
A_7	[3094, 4368] = <3731, 637>	1	→	3	→	12	→	40	→	41	→	
			[205, 313]		[552, 652]		[691, 937]		[130, 194]		[517, 775]	
		55	→	63	→	70						
			[722, 1082]		[277, 415]							
A_8	[3209, 4365] = <3787, 578>	1	→	3	→	12	→	23	→	33	→	
			[205, 313]		[552, 652]		[313, 419]		[232, 348]		[502, 702]	
		46	→	56	→	58	→	62	→	70		
			[249, 373]		[277, 415]		[485, 627]		[394, 516]			
A_9	[3286, 4360] = <3823, 537>	1	→	3	→	2	→	5	→	26	→	
			[205, 313		[87, 131]		[134, 200]		[770, 930]		[198, 296]	
		34	→	47	→	48	→	57	→	62	→	70
			[126, 190]		[308, 462]		[656, 760]		[408, 562]		[394, 516]	

Table 5.5: Nine paths from Boston to Los Angeles as in Fig. 5.4 and their lengths (Okada & Gen (1994))

	A_1	A_2	A_3	A_4	A_5	A_6	A_7	A_8	A_9	A_3	A_4	A_2	A_5	A_6	A_7	A_1	A_8	A_9
A_1	$\mathcal{T}_1$ .500	$\mathcal{T}_2$ .828	$\mathcal{T}_2$ .639	$\mathcal{T}_2$ .625	$\mathcal{T}_2$ .379	$\mathcal{T}_2$ .372	$\mathcal{T}_2$ .376	$\mathcal{T}_2$ .290	$\mathcal{T}_2$ .264	4.273	3.634	3.009	2.181	1.802	1.430	1.054		
A_2	.172	$\mathcal{T}_1$ .500	$\mathcal{T}_2$ .512	$\mathcal{T}_2$ .500	$\mathcal{T}_2$ .261	$\mathcal{T}_2$ .268	$\mathcal{T}_2$ .279	$\mathcal{T}_2$ .210	$\mathcal{T}_2$ .195	2.897	2.385	1.885						
A_3	.361	.488	$\mathcal{T}_1$ .500	$\mathcal{T}_2$ .286	$\mathcal{T}_2$ .076	$\mathcal{T}_2$ .123	$\mathcal{T}_2$ .154	$\mathcal{T}_2$ .113	$\mathcal{T}_2$ .115	2.216								
A_4	.375	.500	.714	$\mathcal{T}_1$ .500	$\mathcal{T}_2$ .050	$\mathcal{T}_2$ .107	$\mathcal{T}_2$ .143	$\mathcal{T}_2$ .105	$\mathcal{T}_2$ .109	2.603	1.889							
A_5	.621	.739	.924	.950	$\mathcal{T}_1$ .500	$\mathcal{T}_2$ .313	$\mathcal{T}_2$ .360	$\mathcal{T}_2$ .143	$\mathcal{T}_2$ .136	4.686	3.762	2.812	2.073					
A_6	.628	.732	.877	.893	.687	$\mathcal{T}_1$ .500	$\mathcal{T}_2$ .444	$\mathcal{T}_2$ .103	$\mathcal{T}_2$ .110	4.974	4.097	3.204	2.472	1.785				
A_7	.624	.721	.846	.857	.640	.556	$\mathcal{T}_1$ .500	$\mathcal{T}_2$ .050	$\mathcal{T}_2$ .080	4.874	4.028	3.171	2.450	1.810	1.254			
A_8	.710	.790	.887	.895	.857	.897	.950	$\mathcal{T}_1$ .500	$\mathcal{T}_2$ .122	6.608	5.721	4.826	4.036	3.179	2.282	1.332	.622	
A_9	.736	.805	.885	.891	.864	.890	.920	.878	$\mathcal{T}_1$ .500	7.369	6.484	5.593	4.788	3.924	3.034	2.114	1.378	.500

Table 4.6: The final table (for $p = 1$) for the problem referred in Section 5.4.

5.6. References

H. **Ishibuchi** & H. **Tanaka (1990)**, Multiobjective programming in optimization of the interval objective function, *European Journal of Operational Research* 48 (2): 219 –225.

S. **Okada** & M. **Gen (1994)**, Order Relation between Intervals and Its Application to Shortest Path Problem, *Japanese journal of Fuzzy Theory and Systems* 6 (6): 703 – 717 (In English).

H. **Schjaer-Jacobsen (1996)**, A new method for evaluating worst- and best-case (WBC) economic consequence of technological development, *International Journal of Production Economics* 46-47: 241 – 250.

H. **Schjaer-Jacobsen (2002)**, Representation and calculation of economic uncertainties: Intervals, fuzzy numbers and probabilities, *International Journal of Production Economics* 78: 91 – 98.

A. **Sengupta** & T.K. **Pal (2006)**, Solving Shortest Path Problem with Interval Arcs, *Fuzzy Optimization and Decision Making* 5(1): 71 - 89.

Chapter 6

Travelling Salesman Problem with Interval Cost Constraints

6.1. Introduction

The Travelling Salesman Problem (TSP) is one of the classical discrete (combinatorial) optimization problems, encountered in Operations Research (Lawler et al (1985), Clifford & Siu (1995)). TSP is also one of such problems considered as puzzles by mathematicians. Suppose a salesman has to visit n cities (or nodes) cyclically and in one tour he has to visit each city just once, and finish up where he started. In what order should he visit them to minimize the cost or distance or time travelled? Much of the work on the TSP is not motivated by direct applications, but rather by the fact that the TSP provides an ideal platform for the study of general methods that can be applied to a wide range of discrete optimization problems (Grötschel & Holland (1991), Reinelt (1994), Applegate et al (1998), Cook (web document), Hoffman (web document)).

As another application of the proposed interval ranking method stated in Chapter 3, the present chapter describes a methodology to solve a Travelling Salesman Problem using interval arcs. In this chapter, an Interval-valued Travelling Salesman Problem (ITSP) is framed and the *cost* or *distance* or *time* of journey between every pair of cities are considered as interval numbers as these parameters may

A. Sengupta and T.K. Pal: Fuzzy Preference Ordering, STUDFUZZ 238, pp. 111–119.
springerlink.com

fluctuate considerably depending on traffic condition, payload etc. The problem defined here is said to be an asymmetrical one because of the cost (or distance or time etc.) between each pair of cities is considered independent of the direction of the journey. The objective of this problem is to select such a route that starts from his home city, passes through each city once and only once, and returns to his home city at lowest possible cost (or in least time or in the shortest possible distance).

6.1.1. *The problem*

A medical representative has to visit five stations A_1 to A_5. He wishes to start from his home place A_1 and wishes to return to his starting station after covering all stations. He does not want to visit any station twice before completing the tour of all stations. Interval costs (given in $A = \langle m(A), w(A) \rangle$ structure) of going from one station to another station are given in Table 6.1. In particular, as it is not possible to go from A_1 to A_1 itself and so on, cells on the principal diagonal alignment conventionally contain ∞ (infinity). The problem here is to determine the optimal route for travelling.

From \ To	A_1	A_2	A_3	A_4	A_5
A_1	∞	$\langle 130, 20 \rangle$	$\langle 135, 10 \rangle$	$\langle 140, 5 \rangle$	$\langle 120, 30 \rangle$
A_2	$\langle 135, 15 \rangle$	∞	$\langle 130, 20 \rangle$	$\langle 130, 15 \rangle$	$\langle 100, 10 \rangle$
A_3	$\langle 60, 5 \rangle$	$\langle 60, 10 \rangle$	∞	$\langle 60, 5 \rangle$	$\langle 75, 10 \rangle$
A_4	$\langle 90, 20 \rangle$	$\langle 100, 20 \rangle$	$\langle 110, 5 \rangle$	∞	$\langle 105, 10 \rangle$
A_5	$\langle 50, 10 \rangle$	$\langle 50, 10 \rangle$	$\langle 55, 5 \rangle$	$\langle 50, 15 \rangle$	∞

Table 6.1: An Interval Traveling Salesman Problem

6.2. Algorithm for Interval-valued Traveling Salesman Problem (ITSP)

This algorithm is based on traditional assignment technique. For interval data the primary requirement is to get a suitable interval ordering methodology because most of the methodologies have there own inherent limitations in ordering any two or a set of interval numbers, particularly when some of them are nested intervals.

Here we use *acceptability index* and *fuzzy preference order* simultaneously in different phases of analysis. Both of a value-based ranking index and a preference ranking index for interval attributes have some specific role in real decision making processes. In an uncertain environment, one can not be replaced universally by the other. Here in this process one may find such a space for using both of the interval ranking methodologies. The algorithm is as stated below.

6.2.1. In order to get the reduced matrix,

6.2.1.1. find the *smallest interval value* (SIV) (using the theory described in Section 3.2) of each row of the cost / distance matrix, then

6.2.1.2. subtract SIV from all the interval elements of the respective rows.

6.2.2. Repeat the same for all columns, i.e., modify the resulting matrix by subtracting the minimum element of each column from all the elements of the respective elements.

6.2.3. Draw a minimum number of horizontal and/or vertical lines to connect all cells containing *zero-centered* intervals.

6.2.3.1. *If* the number of horizontal and/or vertical lines is less than the total number of rows / columns,

then select the minimum from the uncovered elements of the resultant matrix.

And Subtract the minimum value from all uncovered elements and add the same element at the intersection of horizontal and vertical lines.

Thus the next modified matrix is obtained.

6.2.3.2. *If* the number of lines is equal to the total number of rows / columns

then optimal situation is achieved and go to *Step 6.2.4*,

Otherwise go to *Step 6.2.3.3.*

6.2.3.3. Repeat *Step 6.2.3*, until the number of lines is equal to the total number of rows / columns.

6.2.4. Select the '*zero centered cells*' in such a way that a row-wise a single zero is found, mark this cell with the box '□', to make the assignment. Here it is to be noted that if the row / column contains more than one zero-centered-cells assignment should be at the zero valued cell with minimum uncertainty. Then, mark a cross '×' over all zeros if lying in the column of the marked '□', which implies that no assignment can be considered in future there. Continue the process for all zeros and repeat the same for all columns.

Ultimately, in the optimal assignment only one assignment '□' will be there in each row and each column of the matrix.

6.2.5. If the *optimal assignment* satisfies the *additional restriction* (i.e., the salesperson does not want to visit twice or skip any station before completing the tour of all stations and be back to his home station) of the ITSP, then the salesman's optimal path is achieved.

But, if the *additional restriction* is violated, then examine the matrix for some 'next best' solution of the assignment problem. In the selection of the 'next best' cell, there may exist other zero-valued cell with greater uncertainty or any cell containing next minimum interval. And repeat the *Step 6.2.5* until the *additional restrictions* of the ITSP are satisfied.

Optimal assignment may find multiple of paths through the cells having zero-centered intervals. So, optimality in Kundu(1997)'s way or Wang et al (2005)'s way of ranking intervals or, the optimality through $\mathscr{A}$-index is to be predestined. In order to satisfy the additional restriction and to accommodate *next best solution* from an interval-valued data set a preference ranking methodology may be adopted. Advantage of using fuzzy preference ordering scheme as discussed earlier may help in complex situations.

6.3. **Solution to the numerical example 6.1.1**

6.3.1. The Row operation:

6.3.1.1. SIVs of each row of the cost / distance matrix are computed and are shown in the shaded box of Table 6.2.

6.3.1.2. Row SIVs are subtracted from all the interval elements of the respective rows and we get Table 6.3.

6.3.2. Alike operations are repeated for all columns. Shaded boxes of Table 5.3 are the column-wise SIVs. Now, by subtracting the SIV of each column from all the elements of the respective columns, we get Table 6.4.

6.3.3. Minimum numbers of horizontal and/or vertical lines are drawn to connect all cells containing zero-centered intervals. Only *four* lines can only be drawn (Table 6.4).

To

From	A_1	A_2	A_3	A_4	A_5
A_1	∞	$\langle 130, 20\rangle$	$\langle 135, 10\rangle$	$\langle 140, 5\rangle$	**$\langle 120, 30\rangle$**
A_2	$\langle 135, 15\rangle$	∞	$\langle 130, 20\rangle$	$\langle 130, 15\rangle$	**$\langle 100, 10\rangle$**
A_3	**$\langle 60, 5\rangle$**	$\langle 60, 10\rangle$	∞	**$\langle 60, 5\rangle$**	$\langle 75, 10\rangle$
A_4	**$\langle 90, 20\rangle$**	$\langle 100, 20\rangle$	$\langle 110, 5\rangle$	∞	$\langle 105, 10\rangle$
A_5	**$\langle 50, 10\rangle$**	**$\langle 50, 10\rangle$**	$\langle 55, 5\rangle$	$\langle 50, 15\rangle$	∞

Table 6.2: Row-wise SIVs before the Row Operation

To

From	A_1	A_2	A_3	A_4	A_5
A_1	∞	$\langle 10, 50\rangle$	$\langle 15, 40\rangle$	$\langle 20, 35\rangle$	$\langle 0, 60\rangle$
A_2	$\langle 35, 25\rangle$	∞	$\langle 30, 30\rangle$	$\langle 30, 25\rangle$	**$\langle 0, 20\rangle$**
A_3	**$\langle 0, 10\rangle$**	**$\langle 0, 15\rangle$**	∞	**$\langle 0, 10\rangle$**	$\langle 15, 15\rangle$
A_4	$\langle 0, 40\rangle$	$\langle 10, 40\rangle$	$\langle 20, 25\rangle$	∞	$\langle 15, 30\rangle$
A_5	$\langle 0, 20\rangle$	$\langle 0, 20\rangle$	**$\langle 5, 15\rangle$**	$\langle 0, 25\rangle$	∞

Table 6.3: Row-wise SIVs after the Row Operation

To

From	A_1	A_2	A_3	A_4	A_5
A_1	∞	$\langle 10, 65\rangle$	$\langle 10, 55\rangle$	$\langle 20, 45\rangle$	$\langle 0, 80\rangle$
A_2	$\langle 35, 35\rangle$	∞	$\langle 25, 45\rangle$	$\langle 30, 35\rangle$	$\langle 0, 40\rangle$
A_3	$\langle 0, 20\rangle$	$\langle 0, 30\rangle$	∞	$\langle 0, 20\rangle$	$\langle 15, 35\rangle$
A_4	$\langle 0, 50\rangle$	**$\langle 10, 55\rangle$**	$\langle 15, 40\rangle$	∞	$\langle 15, 50\rangle$
A_5	$\langle 0, 30\rangle$	$\langle 0, 35\rangle$	$\langle 0, 30\rangle$	$\langle 0, 35\rangle$	∞

Table 6.4: SIV among uncovered cells after the Column Operation

6.3.4. As numbers of drawn lines are less than the total number of rows / columns, so optimality condition has not been achieved and we find

$\langle 10, 55 \rangle$ as SIV from among the uncovered elements from the matrix in Table 6.4. Now, the SIV $\langle 10, 55 \rangle$ is subtracted from all uncovered elements and added with the elements in cells containing intersecting horizontal and vertical lines. Thus the next modified matrix is obtained (Table 6.5).

6.3.5. In Table 6.5, minimum number of drawn lines connecting all zero-centered cells is equal to the number of rows / columns, so here the optimality condition has been achieved. Now, the solution with the assignment of zeros arrived to anticipated optimality in Table 6.6. This optimality condition may be called the *zero-centered interval optimality*. The following Fig. 6.1 and Tables 6.7 & 6.8 illustrate that there are only two optimal traveling routes that assign one zero-centered cell in each row/column as well as satisfy the *additional restriction* of the TSP.

To

From	A_1	A_2	A_3	A_4	A_5
A_1	∞	$\langle 0, 120 \rangle$	$\langle 0, 110 \rangle$	$\langle 10, 100 \rangle$	$\langle 0, 80 \rangle$
A_2	$\langle 35, 35 \rangle$	∞	$\langle 15, 100 \rangle$	$\langle 20, 90 \rangle$	$\langle 0, 40 \rangle$
A_3	$\langle 10, 75 \rangle$	$\langle 0, 30 \rangle$	∞	$\langle 0, 20 \rangle$	$\langle 25, 90 \rangle$
A_4	$\langle 0, 50 \rangle$	$\langle 0, 110 \rangle$	$\langle 5, 95 \rangle$	∞	$\langle 15, 50 \rangle$
A_5	$\langle 10, 85 \rangle$	$\langle 0, 35 \rangle$	$\langle 0, 30 \rangle$	$\langle 0, 35 \rangle$	∞

Table 6.5: Optimality test.

To

From	A_1	A_2	A_3	A_4	A_5
A_1	∞	$\langle 0, 120 \rangle$	$\langle 0, 110 \rangle$	$\langle 10, 100 \rangle$	$\langle 0, 80 \rangle$
A_2	$\langle 35, 35 \rangle$	∞	$\langle 15, 100 \rangle$	$\langle 20, 90 \rangle$	$\langle 0, 40 \rangle$
A_3	$\langle 10, 75 \rangle$	$\langle 0, 30 \rangle$	∞	$\langle 0, 20 \rangle$	$\langle 25, 90 \rangle$
A_4	$\langle 0, 50 \rangle$	$\langle 0, 110 \rangle$	$\langle 5, 95 \rangle$	∞	$\langle 15, 50 \rangle$
A_5	$\langle 10, 85 \rangle$	$\langle 0, 35 \rangle$	$\langle 0, 30 \rangle$	$\langle 0, 35 \rangle$	∞

Table 6.6: Solution of the ITSP with *zero-centered interval optimality*

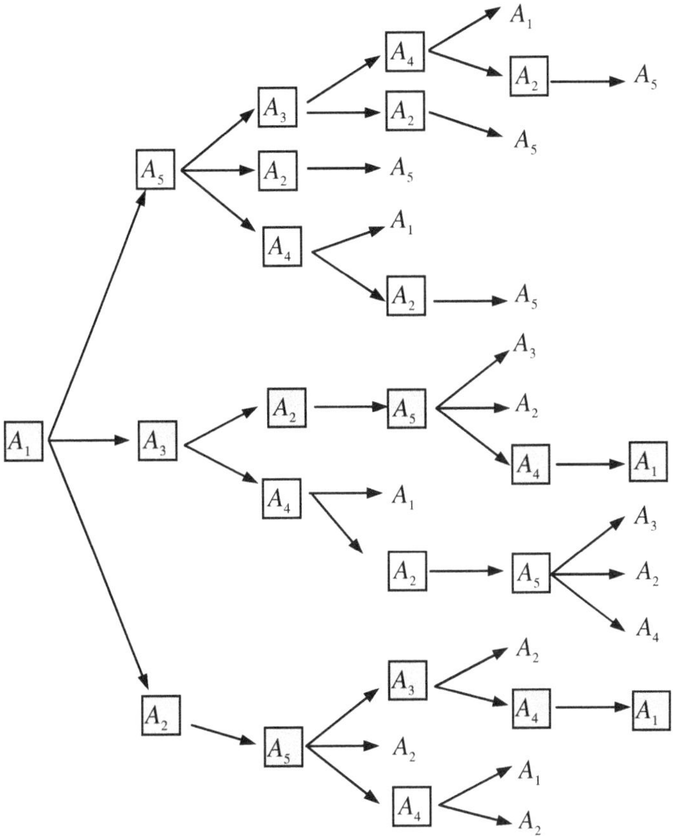

Fig. 6.1. Of all possible routes under zero-centered optimality, only two optimal routes (denoted by shaded boxes) satisfy the TSP constraints.

From \ To	A_1	A_2	A_3	A_4	A_5
A_1	∞	$\langle 0, 120\rangle$	$\langle 0, 110\rangle$	$\langle 10, 100\rangle$	$\langle 0, 80\rangle$
A_2	$\langle 35, 35\rangle$	∞	$\langle 15, 100\rangle$	$\langle 20, 90\rangle$	$\langle 0, 40\rangle$
A_3	$\langle 10, 75\rangle$	$\langle 0, 30\rangle$	∞	$\langle 0, 20\rangle$	$\langle 25, 90\rangle$
A_4	$\langle 0, 50\rangle$	$\langle 0, 110\rangle$	$\langle 5, 95\rangle$	∞	$\langle 15, 50\rangle$
A_5	$\langle 10, 85\rangle$	$\langle 0, 35\rangle$	$\langle 0, 30\rangle$	$\langle 0, 35\rangle$	∞

Table 6.7: 1st *zero-centered interval Optimum route of* the ITSP.

From \ To	A_1	A_2	A_3	A_4	A_5
A_1	∞	$\langle 0, 120 \rangle$	**$\langle 0, 110 \rangle$**	$\langle 10, 100 \rangle$	$\langle 0, 80 \rangle$
A_2	$\langle 35, 35 \rangle$	∞	$\langle 15, 100 \rangle$	$\langle 20, 90 \rangle$	**$\langle 0, 40 \rangle$**
A_3	$\langle 10, 75 \rangle$	**$\langle 0, 30 \rangle$**	∞	$\langle 0, 20 \rangle$	$\langle 25, 90 \rangle$
A_4	**$\langle 0, 50 \rangle$**	$\langle 0, 110 \rangle$	$\langle 5, 95 \rangle$	∞	$\langle 15, 50 \rangle$
A_5	$\langle 10, 85 \rangle$	$\langle 0, 35 \rangle$	$\langle 0, 30 \rangle$	**$\langle 0, 35 \rangle$**	∞

Table 6.8: 2nd *zero-centered interval Optimum route of* the ITSP.

6.3.6. Two non-inferior ITSP routes are obtained as

(*i*) $A_1 \rightarrow A_2 \rightarrow A_5 \rightarrow A_3 \rightarrow A_4 \rightarrow A_1$ (Table 5.7) and here the minimum total cost of travelling is $\langle 435, 60 \rangle$.

(*ii*) $A_1 \rightarrow A_3 \rightarrow A_2 \rightarrow A_5 \rightarrow A_4 \rightarrow A_1$ (Table 5.8) and here the minimum total cost of travelling is $\langle 435, 65 \rangle$.

Hence we need a preference judgment to choose between the two. Fuzzy preference ordering scheme selects the 1st route as the best solution or having preferred minimum total cost of travelling in the above ITSP.

6.4. Conclusion

This chapter describes a TSP using interval arcs. The salesman here wants to visit a certain number of cities allotted to him, where the *cost* or *distance* or *time* of journey between every pair of cities are considered as interval numbers as these may fluctuate considerably depending on traffic condition, payload etc.

The objective of this work in narrower sense is to select such a route that starts from the salesman's home city, passes through each city allotted to him once and only once, and returns to his home city at lowest possible cost (or in least time or in the shortest possible distance). However, the objective in wider sense is to find another application of interval parameters in decision problems and to define a specific choice between existing interval order relations. Hence, both of the value-based ranking index and the preference ranking index for interval attributes have been used. For finding SIVs, the acceptability index is used and for selecting the best interval path from the zero-centered interval optimal paths, a preference ranking methodology is used. In this chapter we have considered a simple ITSP. In complex decision situation, when optimal and near-optimal solutions are in plenty, then fuzzy preference ordering with its totality may be regarded the most suitable one.

In spite of some direct importance of the ITSP problem, quite a lot of practical applications of discrete optimization can be put in this form. So, in future, development of algorithm for large and complex problems is necessary.

6.5. References

D. **Applegate**, R.E. Bixby, V. Chvatal & W. Cook **(1998)**, On the solution of traveling salesman problems, *Documenta Mathematica - Extra Volume, ICM – III*: 645-658.

S.T. **Clifford** & W.C. **Siu (1995)**, New approach for solving the travelling salesman problem using self-organizing learning, in: *Proc. of the IEEE International Conference on Neural Networks*, 5: 2632–2635.

William J. **Cook (web document)**, The Traveling Salesman Problem, (Available online: http://www.tsp.gatech.edu//apps/index.html).

M. **Grötschel** & O. **Holland (1991)**, Solution of Large Scale Symmetric Traveling Salesman Problems, *Mathematical Programming* 51: 141-202.

Karla **Hoffman (web document)**, Traveling Salesman Problem, (Available online: http://iris.gmu.edu/~khoffman/papers/trav_salesman.html).

S. **Kundu (1997)**, Min-transitivity of fuzzy leftness relationship and its application to decision making, *Fuzzy Sets and Systems* 86 (3): 357–367.

E.L. **Lawler**, JK. Lenstra, A.H.G. Rinnooy Kan & D.B. Shmoys **(1985)**, *The Traveling Salesman Problem*, Wiley, New York.

G. **Reinelt (1994)**, *The Traveling Salesman: Computational Solutions for TSP Applications*, Springer-Verlag, Berlin.

Y.M. **Wang**, J.B. Yang & D.L. Xu **(2005)**, A preference aggregation method through the estimation of utility intervals *Computers & Operations Research* 32: 2027–2049.

Chapter 7

Interval Transportation Problem with Multiple Penalty Factors

7.1. Introduction

The problems of distribution of goods from manufacturer to customer are generally described under a common heading, *Transportation Problem* (TP). The TP, originally developed by Hitchcock (1941), can be used when a firm tries to decide where to locate a new facility. Good financial decisions concerning facility location also attempt to minimize total transportation and production costs for the entire system. Moreover, there are many problems not exactly being called the TP but can be modelled alike.

The TP always follows a particular mathematical structure in its constraint formation. Source parameters (a_i) may be the production units, facilities, primary warehouses, etc., and the destination parameters (b_i) may comprise intermediate or secondary warehouses, sales outlets, etc. The penalty factors (c_{ij}) or the coefficients of the objective functions may represent transportation cost, average delivery time of the commodities, unfulfilled demand and so forth.

Several researchers have made intensive investigations on Multi-objective Linear Transportation Problems (MLTP). Aneja & Nair (1979) presented a bi-criteria transportation model. Isermann (1986) proposed an algorithm that generated a set

A. Sengupta and T.K. Pal: Fuzzy Preference Ordering, STUDFUZZ 238, pp. 121–137.
springerlink.com

of all non-dominated solutions to an MLTP. Current & Min (1986) and Current & Marsh (1993) reviewed various aspects of multi-objective design of transportation networks. Climaco et al. (1993) and Ringuest & Rinks (1987) developed two interactive algorithms for the MLTP.

Various efficient algorithms were developed for solving transportation problems with the assumption of precise source, destination parameter, and the penalty factors. However, in reality, these conditions may not be satisfied always. To deal with inexact coefficients in mathematical programming, fuzzy and interval programming techniques have been proposed in various ways (Soyster (1979), Bitran (1980), Steuer (1981), Asai & Tanaka (1984), Ishibuchi & Tanaka (1990), Inuiguchi & Kume (1991), Okada & Gen (1994), Tong (1994), Chanas & Kuchta (1996a, 1996b, 1998), Chinneck & Ramadan (2000), Yorke-Smith & Gervet (2001), Giove (2002), Karmakar & Mujumdar (2005), Oliveira & Antunes (2007) etc.).

In this chapter, we propose a methodology for solving an Interval-valued Transportation Problem where multiple but a synchronized set of penalty factors is involved. The main lacuna in developing a methodology for dealing an Interval-valued Programming Problem possibly lays in the absence of a proper interval order. Therefore, so-called *fuzzy techniques* have been used for solving even a single objective interval-valued programming problem (Das et al (1999)). In continuation with the works stated in previous chapters, here we propose a methodology that uses Interval Arithmetic and existing Interval Order relations.

7.2. Problem formulation

Interval-valued Transportation Problem with Multiple Penalty Factors (ITPMPF) is the problem of minimizing K interval-valued objective functions with both interval source and destination parameters, which can be stated as follows:

$$\text{Minimize } Z^k = \sum_{i=1}^{m}\sum_{j=1}^{n}\left[c_{\text{L}ij}^k,\ c_{\text{R}ij}^k\right] x_{ij} \qquad \text{where } k = 1, 2, \ldots., K,$$

subject to,

$$\sum_{j=1}^{n} x_{ij} = [a_{\text{L}i},\ a_{\text{R}i}], \qquad i = 1, 2, \ldots., m,$$

$$\sum_{i=1}^{m} x_{ij} = [b_{\text{L}j},\ b_{\text{R}j}], \qquad j = 1, 2, \ldots., n,$$

$$x_{ij} \geq 0, \qquad i = 1, 2, \ldots, m; \quad j = 1, 2, \ldots, n,$$

with

$$\sum_{i=1}^{m} \tfrac{1}{2}\left(a_{\mathrm{L}i} + a_{\mathrm{R}i}\right) = \sum_{j=1}^{n} \tfrac{1}{2}\left(b_{\mathrm{L}j} + b_{\mathrm{R}j}\right)$$

where, $[c_{\mathrm{L}ij}^{k}, c_{\mathrm{R}ij}^{k}]$, $(k = 1, 2 \ldots K)$ is an interval representing the inexact cost components for the transportation problem; it can be transportation cost, delivery time, unfulfilled demand and many others. Source and destination parameters respectively lie within the intervals $[a_{\mathrm{L}i}, a_{\mathrm{R}i}]$ and $[b_{\mathrm{L}j}, b_{\mathrm{R}j}]$, and like the conventional one we assume that this is a balanced transportation problem. However, for a balanced problem, $[a_{\mathrm{L}i}, a_{\mathrm{R}i}]$ and $[b_{\mathrm{L}j}, b_{\mathrm{R}j}]$ need not simply be assumed to be identical to each other, instead, these parameters are required to be non-inferior to one another, i.e., one interval parameter may be nested in another but they should be equi-centered.

However, for solution, the techniques of linear programming cannot be applied if and unless the above interval-valued structure of the problem is reduced into a standard linear programming structure and for this purpose we have to resolve the following main issues:

- First, regarding interpretation and realization of the interval-valued order relations.
- Second, regarding interpretation and realization of the decision-making goal, '*Minimize* Z^k' with respect to an inexact environment.

In this chapter, mainly we concentrate on a satisfactory solution approach based on DM's interpretation and realization of ranking of interval numbers and the goal the problem with respect to the inexact environment.

7.3. The scope of an Interval-valued Objective Function in a real decision set up

Earlier in Chapters 2 – 4, we have studied in detail the need for an efficient interval-ordering scheme for any two interval-numbers. Those features guide the present analysis.

The objective of a *conventional* linear programming problem (LPP) is to maximize or minimize the value of its (one only, single-valued) objective function satisfying a given set of restrictions. However, a single-objective interval linear programming problem (ILPP) contains an interval-valued objective function (IOF). Let us consider the following problem:

$$\text{Maximize / Minimize } Z = \sum_{j=1}^{n} [c_{Lj},\ c_{Lj}]x_j \tag{7.3.1}$$

subject to,

$$\{\text{A set of feasibility constraints}\} \tag{7.3.2}$$

An interval can be represented by any two of its four attributes (viz., left limit, right limit, mid-value and width) (Ishibuchi & Tanaka (1990)). However, in our study of comparing interval numbers as stated in the earlier chapters, we have given much importance on two attributes, the mid-value and the width of the intervals, than the other two, the left and the right limit. In search of more precision, here also for a maximization / minimization problem, the minimization of width along with maximization / minimization of the mid-value of the IOF carries much more significance to our intuition. Therefore, by using the attributes, the mid-value and width, the ILPP, (7.3.1)-(7.3.2), is reduced into a bi-objective LPP as follows:

$$\text{Max / Min } \{\textit{mid-value} \text{ of the IOF}\} \tag{7.3.3}$$

$$\text{Min } \{\textit{width} \text{ of the IOF}\} \tag{7.3.4}$$

subject to

$$\{\text{set of feasibility constraints}\}. \tag{7.3.5}$$

From this problem, naturally one may get two conflicting optimal solutions:

$$x' = \{x'_j\}, \text{ from (7.3.3) and (7.3.5)}$$

$$x'' = \{x''_j\}, \text{ from (7.3.4) and (7.3.5)}$$

and hence we get two optimal values Z' and Z'' respectively.

If $x' = x''$, then there does not exist any conflict and x' is the solution of the problem. But, for the maximization problem,

$$\text{if } x' \neq x'', \text{ then } m(Z') > m(Z'') \text{ and } w(Z') > w(Z''),$$

(because, Z' is obtained through maximizing $m(Z)$ and Z'' is obtained through the other goal, by minimizing $w(Z)$). Similarly, for the minimizing problem,

$$\text{if } x' \neq x'', \text{ then } m(Z') < m(Z'') \text{ and } w(Z') > w(Z''),$$

(because Z' here is obtained by minimizing $m(Z)$ and Z'' by minimizing $w(Z)$). Therefore, if $x' \neq x''$, then Z' and Z'' become the non-dominated extreme alternatives (Ignizio (1982)).

On the other hand, the principle of $\mathscr{A}$-*index* indicates that for the maximization (minimization) problem, an interval with a higher mid-value is superior (inferior) to an interval with a lower mid-value. Therefore, though Z' and Z'' are two non-dominated alternative extremes from the viewpoint of a bi-objective problem, they can be ranked through $\mathscr{A}$-*index*.

Hence, in order to obtain *maximum / minimum* of the interval objective function, considering the mid-value of an interval-valued objective function is our primary concern. Therefore, if we reduce the interval objective function in its central value and use conventional LP techniques for favour of its solution, the solution will give the best-expected optimum for the problem concerned. Further, we also need to consider the width but as a secondary attribute, only to determine best reachable certainty level and to confirm whether the best-expected optimum is within the acceptable limit of the DM for the problem concerned. If it is not, one has to go for smaller extent of width (uncertainty) according to his satisfaction and thus to obtain a less wide interval from among the non-dominated alternatives accordingly.

Henceforth, taking into account (7.3.3) and (7.3.4), we develop a composite goal to define the IOF as follows:

For maximization problem:

$$\text{Maximize } \lambda.m(Z) + (1-\lambda)(-w(Z)) \tag{7.3.6}$$

subject to,

(7.3.5) and $0 \leq \lambda \leq 1$.

For minimization problem:

$$\text{Minimize } \lambda.m(Z) + (1-\lambda)w(Z) \tag{7.3.7}$$

subject to,

(7.3.5) and $0 \leq \lambda \leq 1$.

The lambda (λ) factor defines the DM's pessimistic or optimistic bias. If $\lambda = 1$, the problems (7.3.6) & (7.3.7) show DM's absolute optimistic bias and on the contrary, if $\lambda = 0$, then the problems (7.3.6) & (7.3.7) indicate the pessimistic DM's attitude. With $\lambda = .5$, or any value close to .5, a similar proportional balance between DM's optimistic and pessimistic preference is apparent.

From this juncture, let us move towards a Multi-objective Interval-valued Linear Programming Problem with a like structure to (7.3.1) & (7.3.2), however with a multiplicity of Z, such that

$$Z = \left\{ Z^i \mid i = 1,\ 2,\ \ldots,\ m \right\}.$$

The composite OF for the *maximization problem* is then reduced to,

$$\text{Maximize } \lambda \sum_{i=1}^{m} v_i m(Z^i) - (1-\lambda) \sum_{i=1}^{m} v_i w(Z^i) \tag{7.3.8}$$

and similarly, the composite OF for the *minimization problem* is,

$$\text{Minimize } \lambda \sum_{i=1}^{m} v_i m(Z^i) + (1-\lambda) \sum_{i=1}^{m} v_i w(Z^i) \tag{7.3.9}$$

where, weights v_i ($\forall i$ = 1, 2, …., m), such that $\sum_{i=1}^{m} v_i = 1$, are attached to the OFs to facilitate the DM with more control over the decision-making process.

If the λ value is chosen closer to unity, the solution process reflects the DM's inclination more towards optimism. Here the mean value component of the interval OF is given highest preference than the uncertainty component of the OF. The more the value of λ, the more the importance given to the mean of the interval OF. On the contrary, if the λ value is chosen closer to zero, the uncertainty component of the interval OF is given more preference than the interval OF mean and this reflects the DM's pessimistic bias in the solution process. Similarly, if both the components mean and uncertainty of any interval OF are to be given 'nearly equal' preferences, the value of λ comes somewhere close to 0.5, and this reflects the DM's sense of rationality and ability to find a compromise in a conflicting situation. On the other hand the proportion of weights v_i distributed to the OFs signifies the DM's assignment of priorities among the conflicting objectives. The DM here can use separate λ factors for the separate OFs. So, $\lambda_i \in [0,1], \forall i$, can be assigned to the ith OF.

Therefore, for the *maximization problem*, the goal (7.3.8) can be reconstructed as,

$$\text{Maximize } \lambda_i \sum_{i=1}^{m} v_i m(Z^i) - (1-\lambda_i) \sum_{i=1}^{m} v_i w(Z^i), \quad \forall i$$

and similarly, for the *minimization problem*, (7.3.9) can be reformed as,

$$\text{Minimize } \lambda_i \sum_{i=1}^{m} v_i m(Z^i) + (1-\lambda_i) \sum_{i=1}^{m} v_i w(Z^i), \quad \forall i.$$

Assigning suitable priorities (or, weights, v_i) among the multiple objectives and taking the λ factor as $\lambda \in [0,1]$ for the for the entire set of the interval objectives or, the individual λ factors, $\lambda_i \in [0,1], \forall i,$ separately for the ith OF ($\forall i$ = 1, 2, …., m), will aid the DM with more insight into the problem concerned as well as more control over the decision-making process.

7.3.1. *A numerical example*

Let us consider the following problem earlier referred by Chanas & Kuchta (1996b):

$$\begin{aligned} &\text{Maximize } Z = [-20,\ 50]x_1 + [0,\ 10]x_2 \\ &\text{subject to,} \\ &\left.\begin{aligned} 10x_1 + 60x_2 &\le 1080 \\ 10x_1 + 20x_2 &\le 400 \\ 10x_1 + 10x_2 &\le 240 \\ 30x_1 + 10x_2 &\le 420 \\ 40x_1 + 10x_2 &\le 520 \\ x_1, x_2 &\ge 0 \end{aligned}\right\} \Rightarrow x \in X \end{aligned} \tag{7.3.1.1}$$

As stated above, the interval objective function Z for the maximization problem is first reduced into a bi-objective optimization problem

$$\text{Max } m(Z) = 15x_1 + 5x_2 \tag{7.3.1.2}$$

$$\text{Min } w(Z) = 35x_1 + 5x_2\,. \tag{7.3.1.3}$$

Now, using (7.3.1.2) and (7.3.1.3) we construct a composite objective function as follows

$$\begin{aligned} \text{Max } Z' &= \lambda\, m(Z) + (1-\lambda)(-w(Z)) \\ &= (50\lambda - 35)x_1 + (10\lambda - 5)x_2, \end{aligned}$$

subject to,

$$\{\text{the set of constraints, } x \in X\},$$

where, the value of λ ranging between 0 and 1 represents the DM's pessimistic and optimistic attitude. Let us solve the problem for λ = .8, i.e., when the DM gives much importance to the maximization of the mid-value of the interval objective function and comparatively very much lower importance to the certainty of the outcome. The solution is found as

$$(x_1, x_2) = (9,\ 15) \text{ and } \langle m(Z),\ w(Z)\rangle = \langle 210,\ 390\rangle.$$

Let the DM is not satisfied with the solution and wish to reduce the level of uncertainty in the maximal interval objective value. Then he may be interested to get the solution of the problem at a lower value of λ, say, at $\lambda = .6$. The solution is obtained as

$$(x_1, x_2) = (0,\ 18) \text{ and } \langle m(Z),\ w(Z)\rangle = \langle 90,\ 90\rangle.$$

7.3.2. *A discussion on Chanas & Kuchta (1996b)'s approach and a comparative analysis with our approach*

Chanas & Kuchta (1996b) did not consider any separate ranking index but restructured Ishibuchi & Tanaka (1990)'s different approaches into a single scheme using t_0,t_1-cut of the comparing intervals, where t_0 and t_1 are any fixed numbers such that $0 \le t_0 < t_1 \le 1$. Here, t_0,t_1-cut of an interval $A = [a_L,\ a_R]$ is defined as

$$A \big/_{[t_0,t_1]} = [a_L + t_0(a_R - a_L),\ a_L + t_1(a_R - a_L)]$$

and Ishibuchi & Tanaka (1990)'s three approaches are reconstructed as follows:

$$A \le_\beta \big/_{[t_0,t_1]} B \iff A \big/_{[t_0,t_1]} \le_\beta B \big/_{[t_0,t_1]},$$

$$A <_\beta \big/_{[t_0,t_1]} B \iff A \big/_{[t_0,t_1]} <_\beta B \big/_{[t_0,t_1]},$$

where

(i) $A \big/_{[t_0,t_1]}$ and $B \big/_{[t_0,t_1]}$ are the t_0,t_1-cuts of the intervals A and B, and

(ii) $\beta \in \{\beta_1,\ \beta_2,\ \beta_3\}$ is any ordering scheme such that relations $\le_{\beta_1}$, $\le_{\beta_2}$ and $\le_{\beta_3}$ are identical to $\le_{LR}$, $\le_{mw}$ and $\le_{Lm}$ of Ishibuchi & Tanaka (1990) respectively (refer Section 2.3.1 for the definitions of $\le_{LR}$, $\le_{mw}$ and $\le_{Lm}$).

Chanas & Kuchta (1996b) dealt the following interval objective linear programming (IOLP) problem,

$$Z(x) = Z(x_1, x_2, \ldots., x_n) = \sum_{i=1}^{n} C_i x_i \quad \rightarrow \text{ max},$$

subject to,

$$x \in X = \{x \in \Re^n : Ax \leq B, \ \ x \geq 0\},$$

in which the coefficients $C_i = [c_{\mathrm{L}i}, \ c_{\mathrm{R}i}]$ are interval numbers.

Now, based on their approach Chanas & kuchta (1996b) tried to find out all pareto-optimal solutions of the above IOLP problem through solving the following bi-criteria linear programming problem,

$$\sum_{i=1}^{n} [c_{\mathrm{L}i} + t_0(c_{\mathrm{R}i} - c_{\mathrm{L}i})]x_i \rightarrow \text{ max},$$

$$\sum_{i=1}^{n} [c_{\mathrm{L}i} + t_1(c_{\mathrm{R}i} - c_{\mathrm{L}i})]x_i \rightarrow \text{ max},$$

$$x \in X.$$

Subsequently the above bi-objective problem was reduced to a parametric linear programming problem in the following form and was solved for different values of *t*,

$$\sum_{i=1}^{n} [c_{\mathrm{L}i} + t(c_{\mathrm{R}i} - c_{\mathrm{L}i})]x_i \rightarrow \text{ max},$$

$$x \in X, \tag{7.3.2.1}$$

$$t \in (t_0, \ t_1).$$

The authors solved the numerical IOLP problem (7.3.1.1) for different values of $t \in (0, 1)$. To compute the solution they used different values of t in 'an approximate way' (Chanas & Kuchta (1996b)). We refer here the results obtained by them in Table 7.1 for further discussion. Here we supplement only the 3rd column to their original findings.

Regarding the above result here we like to present our views in the following points:

1. Chanas & Kuchta (1996b) only tried to unify three concepts (introduced by Ishibuchi & Tanaka (1990)) on preference relation between intervals. Naturally, in their approach there was nothing new regarding the conce-

tual discrepancies in the problem of ranking between any two interval numbers as is stated in case 3(d) of Section 4.6.

For any fixed value of t	Optimal solution	Corresponding optimal OF value $< m(Z),\ w(Z) >$
$t \in (0,\ .29]$	$x^1 = (0,\ 18)$	$< 90,\ 90 >$
$t = .3$	$x^2 = (6,\ 17)$	$< 175,\ 295 >$
$t \in [.31,\ .33]$	$x^3 = (8,\ 16)$	$< 200,\ 360 >$
$t \in [.34,\ .49]$	$x^4 = (9,\ 15)$	$< 210,\ 390 >$
$t \in [.5,\ .66]$	$x^5 = (10,\ 12)$	$< 210,\ 410 >$
$t \in [.67,\ 1)$	$x^6 = (13,\ 0)$	$< 195,\ 455 >$

Table 7.1: Chanas & Kuchta (1996b)'s 'pareto-optimal' solution set of the IOLP Problem (7.3.1.1)

2. Ishibuchi & Tanaka (1990) used the relation $\leq_{\mathrm{Lm}}$ for solving an IOLP problem and obtained an incomparable (or, pareto-optimal) set at most of two solutions (For example, for this particular numerical problem, the set of solutions should contain $x = \{(0, 0),\ (10, 12)\}$. Here, solution $(10, 12)$ is obtained as optimal solution.). On the other hand, in Chanas & Kuchta (1996b)'s *pareto-optimal set* contains as many as six solutions as well as having following contradictions too.

    - *Contradiction 1*: Solutions x^4 and x^5 cannot be members of a set of equally preferred alternative optimal solutions because of their corresponding optimal interval OF values which are given in the column 3 of the above Table 7.1. It contradicts Ishibuchi & Tanaka (1990)'s concept of $\leq_{\mathrm{mw}}$ relation. Interval $< 210, 390 >$ is certainly a preferred interval to $< 210, 410 >$. In different ways this sort of relation has already been referred and explained in Chapter 2 and Chapter 3.

    - *Contradiction 2*: Similarly, solution x^6 is inferior to all of solutions x^3, x^4 and x^5 by virtue of the preference relation $\leq_{\mathrm{mw}}$ and cannot be a co-member of a set of equally preferred alternative optimal solutions.

3. Chanas & Kuchta (1996b)'s approach generates too many solutions, which initiate another decision-making situation for selection of a precis-

ely preferred solution of an IOLP problem from among many alternatives. For a problem with large numbers of variables and constraints, alternative solutions may then increase considerably.

4. Solution x^6 comes from using any value $t \in [.67,\ 1]$. Therefore, for $t = 1$, problem (7.3.2.1) turns out to be

$$m(Z(x)) + w(Z(x)) \quad \rightarrow \text{max},$$
$$x \in X,$$

in which a part of the objective function, '$w(Z(x)) \rightarrow \max$', does not comply with the DM's goal of having a preferred maximum but with less uncertainty.

In contrast, our approach suggests a clear and convincing decision. (All of the optimal solutions can also be obtained by repeated change in the value of λ. However, there is a question whether it is necessary to know the complete set of solutions for a decision-making problem.) If the DM is not satisfied with the available mean value or with the associated level width or uncertainty of the maximizing interval OF value, he may adjust the λ value accordingly in the final LP form. For this interactive procedure for selection of a preferred alternative, a DM need not be bothered with objective level of λ but should rely on his own subjective realization. He also may take help of the *fuzzy preference ordering* for selecting one from close alternatives as is stated in Chapter 4.

By changing the value of λ with the subjective interaction with the DM, we get the following closest possible satisfactory solutions of the problem (7.3.1.1):

For any fixed value of $\lambda \in [0,\ 1]$	Optimal solution $x^{\lambda} = (x_1^{\lambda},\ x_2^{\lambda})$	Corresponding optimal OF value $< m(Z),\ w(Z) >$
$\lambda = .8$	$x^{.8} = (9,\ 15)$	$< 210,\ 390 >$
$\lambda = .73$	$x^{.73} = (8,\ 16)$	$< 200,\ 360 >$
$\lambda = .71$	$x^{.71} = (6,\ 17)$	$< 175,\ 295 >$
$\lambda = .6$	$x^{.6} = (0,\ 18)$	$< 90,\ 90 >$

Table 7.2: Optimal solution set (in our approach) of the IOLP problem (7.3.1.1)

According to Ishibuchi & Tanaka (1990)'s approach, these four optimal objective function values mentioned in Table 7.2 are members of the incomparable (pareto-optimal) set of intervals. However, using fuzzy preference ordering, definitely these intervals can be ranked and here $<90,\ 90>$ comes out as the preferred maximum (taking $p = 1$). Nevertheless, if the DM desires to have his preferred maximum with higher mean value then he should select value of λ close to unity. The preferred maximum objective value with minimum width is available here at $<90,\ 90>$ which is obtained for $\lambda = .6$, i.e., when minimization of width of the objective function is given some importance.

7.4. A numerical example of ITPMPF

The following Numerical Example is taken from Das et al (1999).

$$\text{Minimize } Z^1 = \sum_{i=1}^{3}\sum_{j=1}^{4} c_{ij}^1 x_{ij}$$

$$\text{Minimize } Z^2 = \sum_{i=1}^{3}\sum_{j=1}^{4} c_{ij}^2 x_{ij}$$

Subject to,

$$\sum_{j=1}^{4} x_{1j} = [7, 9], \quad \sum_{j=1}^{4} x_{2j} = [17, 21], \quad \sum_{j=1}^{4} x_{3j} = [16, 18],$$

$$\sum_{i=1}^{3} x_{i1} = [10, 12], \quad \sum_{i=1}^{3} x_{i2} = [2, 4],$$

$$\sum_{i=1}^{3} x_{i3} = [13, 15], \quad \sum_{i=1}^{3} x_{i4} = [15, 17]$$

$$x_{ij} \geq 0, \quad i= 1,2,3, \ j=1,2,3,4.$$

where,

$$C^1 = (c_{ij}^1)_{3\times 4} = \begin{bmatrix} [1,\ 2] & [1,\ 3] & [5,\ 9] & [4,\ 8] \\ [1,\ 2] & [7,\ 10] & [2,\ 6] & [3,\ 5] \\ [7,\ 9] & [7,\ 11] & [3,\ 5] & [5,\ 7] \end{bmatrix} \text{ and}$$

$$C^2 = (c_{ij}^2)_{3\times 4} = \begin{bmatrix} [3,\ 5] & [2,\ 6] & [2,\ 4] & [1,\ 5] \\ [4,\ 6] & [7,\ 9] & [7,\ 10] & [9,\ 11] \\ [4,\ 8] & [1,\ 3] & [3,\ 6] & [1,\ 2] \end{bmatrix}$$

Solution: Both the IOFs are to be realized in the following way –

$$\text{Objective\_1} = \left\{ \ \min m(Z^1)\ \&\ \min w(Z^1) \right\}$$

$$\text{Objective\_2} = \left\{ \ \min m(Z^2)\ \&\ \min w(Z^2) \right\}$$

subject to,

$$\begin{aligned} &7 \le x_{11} + x_{12} + x_{13} + x_{14} \le 9 \\ &17 \le x_{21} + x_{22} + x_{23} + x_{24} \le 21 \\ &16 \le x_{31} + x_{32} + x_{33} + x_{34} \le 18 \\ &10 \le x_{11} + x_{21} + x_{31} \le 12 \\ &2 \le x_{12} + x_{22} + x_{32} \le 4 \\ &13 \le x_{13} + x_{23} + x_{33} \le 15 \\ &15 \le x_{14} + x_{24} + x_{34} \le 17 \\ &x_{ij} \ge 0, \quad i = 1,\ 2,\ 3., \quad j = 1,\ 2,\ 3,\ 4. \end{aligned} \tag{7.4.1}$$

where, $m(Z^1) = 1.5x_{11} + 2x_{12} + 7x_{13} + 6x_{14} + 1.5x_{21} + 8.5x_{22} + 4x_{23} + 4x_{24} + 8x_{31} + 9x_{32} + 4x_{33} + 6x_{34}$,

$$w(Z^1) = .5x_{11} + x_{12} + 2x_{13} + 2x_{14} + .5x_{21} + 1.5x_{22} + 2x_{23} + x_{24} + x_{31} + 2x_{32} + x_{33} + x_{34},$$

$$m(Z^2) = 4x_{11} + 4x_{12} + 3x_{13} + 3x_{14} + 5x_{21} + 8x_{22} + 8.5x_{23} + 10x_{24} + 6x_{31} + 2x_{32} + 4.5x_{33} + 1.5x_{34}, \text{ and}$$

$$w(Z^2) = x_{11} + 2x_{12} + x_{13} + 2x_{14} + x_{21} + x_{22} + 1.5x_{23} + x_{24} + 2x_{31} + x_{32} + 1.5x_{33} + .5x_{34}.$$

In a Transportation Problem, the feasibility constraints are always equality constraints. So, if the right-hand side of a constraint is interval number, an equality constraint with decision variables in the left side can be written as a deterministic set of constraints as above (Tong (1994), Das et al (1999), Sengupta & Pal (2001, 2003, 2004).

The composite OF of the problem (7.4.1) is in its final form as,

$$Z = \lambda\left(v_1 m(Z^1) + v_2 m(Z^2)\right) + (1-\lambda)\left(v_1 w(Z^1) + v_2 w(Z^2)\right) \tag{7.4.2}$$

Now, assuming a rational DM's specific choice, with $\lambda = .5$ and $v_1 = v_2 = .5$, the composite OF of the problem (7.4.1) is

$$Z = .25m(Z^1) + .25m(Z^2) + .25w(Z^1) + .25w(Z^2) \tag{7.4.3}$$

and we have the final form of the LP problem (7.4.1) as

Minimize OF (7.4.3),

subject to,

{the feasibility constraint set of the problem (7.4.1)}.

We get the optimal solution of the problem as follows:

$$x_{12} = 2,\ x_{13} = 5,\ x_{21} = 10,\ x_{23} = 7,\ x_{33} = 1,\ x_{34} = 15,$$

$$< m(Z^1), w(Z^1) > \ = \ < 176, 47 >,$$

$$< m(Z^2), w(Z^2) > \ = \ < 160, 38.5 >.$$

Now, the DM, say for example, wish to assign more importance to first objective and wish to have more certain result, then with $\lambda = .3,\ v_1 = .6$ and $v_2 = .4$, we have the result

$$x_{12} = 2,\ x_{13} = 5,\ x_{21} = 10,\ x_{24} = 7,\ x_{33} = 8,\ x_{34} = 8,$$

$$< m(Z^1), w(Z^1) > \ = \ < 162, 40 >,$$

$$< m(Z^2), w(Z^2) > \ = \ < 191, 42 >.$$

Even Das et al (1999)'s result can be obtained through the following problem,

Minimize OF (7.4.2),

subject to,

{the feasibility constraint set of (7.4.1)}.

If we retain the LP structure of the problem, a few trial-and-error attempts show that, with $\lambda = .5$, $v_1 = .756$ and $v_2 = .244$, we get their result as

$$< m(Z^1), w(Z^1) > \; = \; < 158.95, 45.95 >,$$

$$< m(Z^2), w(Z^2) > \; = \; < 178.875, 48.985 > .$$

7.5. Conclusion

The aim of this chapter is to define an Interval-valued Transportation Problem with Multiple Penalty Factors (ITPMPF) and describe a procedure for obtaining a satisfactory compromise solution. For any inexact programming problem, before getting into the solution procedure, the first task is to explain the *meaning* of the objective defined in the context of the referred environment. So, in the light of *interval order relations* discussed earlier in Chapters 2 – 4, maximization / minimization of an IOF is interpreted and finally a composite objective function of an ITPMPF is defined. The ITPMPF is reduced into a standard LPP structure and solved.

One distinguishable feature of this work is that it reflects DM's pessimistic or optimistic bias in achieving the compromise solution. The change in the value of *lambda* (λ) from 0 to 1 defines a DM's preference changing from pessimism to optimism. Another feature of this work may be mentioned that for an ITPMPF with k IOFs, only a few steps are sufficient to yield the compromise solution of the problem. Another important aspect of this approach is the flexibility, with which the DM finds more hold into the problem than any other approach. The approach stated in this chapter facilitates the DM to have his desired satisfactory solution by changing the objective weights of the multi-objective problem or the optimism / pessimism parameter in suitable ways.

7.6. References

Y.P. **Aneja** & K.P.K. **Nair (1979)**, Bi-criteria Transportation Problems, *Management Science* 25: 73 – 78.

K. **Asai** & H. **Tanaka (1984)**, Fuzzy linear programming with fuzzy numbers, *Fuzzy Sets and Systems* 13 (1): 1 – 10.

G.R. **Bitran (1980)**, Linear multiobjective problems with interval coefficient, *Management Science* 26: 694 – 706.

S. **Chanas** & D. **Kuchta (1996a)**, A concept of solution of the Transportation Problem with fuzzy cost coefficient, *Fuzzy Sets and Systems* 82 (3): 299 – 305.

S. **Chanas** & D. **Kuchta (1996b)**, Multiobjective programming in optimization of interval objective functions – A generalized approach, *European Journal of Operational Research* 94 (3): 594 – 598.

S. **Chanas** & D. **Kuchta (1998)**, Fuzzy integer transportation problem, *Fuzzy Sets and Systems* 98 (3): 291 – 298.

J.W. **Chinneck** & K. **Ramadan (2000)**, Linear programming with interval coefficients, *Journnal of the Operational Research Society* 51 (2): 209 – 220.

J.N. **Climaco**, C.H. Antunes & M.J. Alves **(1993)**, Interactive decision support for multiobjective transportation problems, *European Journal of Operational Research* 65 (1): 58 – 67.

J. **Current** & M. **Marsh (1993)**, Multiobjective transportation network design and routing problems: Taxonomy and annotation, *European Journal of Operational Research* 65 (1): 4 – 19.

J. **Current** & H. **Min (1986)**, Multiobjective design of transportation networks: Taxonomy and annotation, *European Journal of Operational Research* 26 (2): 187 – 201.

S.K. **Das**, A. Goswami & S.S. Alam **(1999)**, Multiobjective transportation problem with interval cost, source and destination parameters, *European Journal of Operational Research* 117 (1): 100 – 112.

Silvio **Giove (2002)**, Interval TOPSIS for Multicriteria Decision Making, in: M. Marinaro & R. Tagliaferri (Eds.): WIRN VIETRI 2002, *Lecture Notes in Computer Science* 2486, Springer-Verlag Berlin Heidelberg: 56-63.

F.L. **Hitchcock (1941)**, The distribution of a product from several sources to numerous localities, *Journal of Mathematical Physics* 20: 224 – 230.

M. **Inuiguchi** & Y. **Kume (1991)**, Goal Programming problem with Interval Coefficients and Target Intervals, *European Journal of Operational Research* 52 (3): 345 – 360.

H. **Isermann (1979)**, The enumeration of all efficient solutions for a Linear Multiobjective Transportation Problem, *Naval Research Logistic Quarterly* 26: 123 – 139; *Fuzzy Sets of Systems* 18 (1986): 15 – 30.

H. **Ishibuchi** & H. **Tanaka (1990)**, Multiobjective programming in optimization of the interval objective function, *European Journal of Operational Research* 48 (2): 219 –225.

S. **Karmakar** & P.P. **Mujumdar (2005)**, Grey Fuzzy Multiobjective Optimization Model for River Water Quality Management, *Proc. of International Conference on Hydrological Perspectives for Sustainable Development* (HYPESD - 2005): 923-931, IIT Roorkee, India. (Available online: http: //eprints.iisc.ernet.in /archive/ 00003153/01/Karmakar_Mujumdar_ HYPESD_2005.pdf).

S. **Kundu (1997)**, Min-transitivity of fuzzy leftness relationship and its application to decision making, *Fuzzy Sets and Systems* 86 (3): 357–367.

S. **Okada** & M. **Gen (1994)**, Order Relation between Intervals and Its Application to Shortest Path Problem, *Japanese journal of Fuzzy Theory and Systems* 6 (6): 703 – 717 (In English).

C. **Oliveira** & C.H. **Antunes** (2007), Multiple objective linear programming models with interval coefficients – an illustrated overview, *European Journal of Operational Research* 181: 1434–1463.

J.L. **Ringuest** & D.B. **Rinks (1987)**, Interactive solutions for the Linear Multiobjective Transportation Problem, *European Journal of Operations Research* 32 (1): 96 – 106.

A. **Sengupta**, T.K. Pal & D. Chakraborty **(2001)**, Interpretation of inequality constraints involving interval coefficients and a solution to interval linear programming, *Fuzzy Sets and Systems* 119 (1): 129 – 138.

A. **Sengupta** & T.K. **Pal (2003)**, Interval-valued transportation problem with multiple penalty factors, *VU Journal of Physical Sciences* 9: 71 – 81.

A. **Sengupta** & T.K. **Pal (2004)**, Decision Maker's Pessimistic / Optimistic Bias in an Interval-valued Transportation Problem, e-Proceedings of the *First World Congress on Lateral-Computing* (WCLC 2004), Bangalore, India.

A.L. **Soyster (1979)**, Inexact Linear Programming with Generalized Resource Sets, *European Journal of Operational Research* 3: 316-321.

R.E. **Steuer (1981)**, Algorithm for linear programming problems with interval objective function coefficient, *Mathematics of Operations Research* 6: 333 – 348.

S. **Tong (1994)**, Interval number and fuzzy number linear programming, *Fuzzy Sets and Systems* 66: 301–306.

N. **Yorke-Smith** & C. **Gervet (2001)**, Data uncertainty in constraint programming: a non-probabilistic approach, American Association for Artificial Intelligence 2001 Fall Symposium, November 2-4, 2001. (Available online: http: //www-users.cs.york.ac.uk/~tw/fall /Proceedings/yorke.pdf).

Chapter 8

Fuzzy Preference based TOPSIS for Interval Multi-criteria Decision Making

8.1. Introduction

In almost all real-world decision making problems multiplicity of criteria for judging the alternatives is pervasive and for many of such problems, the task of decision making is merely to solve a Multiple Criteria Decision-Making (MCDM) problem, which can concisely be expressed in a matrix format as

$$D = \begin{array}{c} \\ A_1 \\ A_2 \\ \vdots \\ \vdots \\ A_m \end{array} \begin{array}{c} \begin{array}{cccccc} C_1 & C_2 & \cdots & \cdots & C_n \end{array} \\ \begin{bmatrix} x_{11} & x_{12} & \cdots & \cdots & x_{1n} \\ x_{21} & x_{22} & \cdots & \cdots & x_{2n} \\ \vdots & \vdots & \cdots & \cdots & \vdots \\ \vdots & \vdots & \vdots & \vdots & \vdots \\ x_{m1} & x_{m2} & \cdots & \cdots & x_{mn} \end{bmatrix} \end{array} \tag{8.1.1}$$

$$W = \begin{bmatrix} w_1 & w_2 & \cdots & w_n \end{bmatrix}$$

A. Sengupta and T.K. Pal: Fuzzy Preference Ordering, STUDFUZZ 238, pp. 139–154.
springerlink.com

where,

1) $A_1, A_2, \cdots, A_m$ are the possible alternatives among which a DM has to choose from,

2) $C_1, C_2, \cdots, C_n$ are the criteria over which alternative performances are measured,

3) x_{ij} is the rating against alternative A_i with respect to the criterion C_j and

4) w_j is the weight assigned to criterion C_j.

In classical MCDM methods, the problem data given in terms of ratings and weights are all known preciously (Teghem et al (1989), Dyer et al (1992), Chen et al (1992)). The TOPSIS method (the Technique for Order Preference by Similarity to the Ideal Solution (Hwang & Yoon (1981)) is one of the most acceptable approaches for solving classical MCDM problems. TOPSIS defines an index called *similarity* or *relative closeness to the ideal solution* by combining the proximity to the ideal solution and the remoteness from the negative-ideal solution. Hence, a decision-making via TOPSIS approaches to select an alternative with maximum relative similarity to the hypothetical ideal one. However, there are many conditions, under which simple crisp data are not adequate to model real-life situations. Since human judgments and preferences are often vague and generally not representable with exact numerical values, in the literature we have several MCDM approaches, which consider inexact values as the problem data (Bellman & Zadeh (1970), Delgado et al (1992), Herrera et al (1996), Hsu & Chen (1996, 1997), Chen (2000), Chen et al (2005), Abo-Sinna & Amer (2005)).

In the decision-making literature, the approach based on interval analysis allows handling imprecise input data. This approach consists in assuming that the data of a decision-making problem are not well defined but may vary in given intervals. Giove (2002) proposed an interval version of the traditional TOPSIS method. This chapter describes I-TOPSIS (Interval TOPSIS) approach using the *acceptability index* introduced in Chapter 3 for comparing crisp interval numbers. The main weakness of *acceptability index* is that, as it is only a value-based ranking index, it can be applicable only partially, for selection of the best alternative from a pessimistic (or a risk averse) DM's point of view. So, in a broader view, enhancing the applicability of *acceptability index* as a pessimistic DM's preference index too and for attaining more flexibility and efficiency of DM's interval ranking strategy, *Fuzzy Preference* in interval ordering has been proposed as a generalized ranking index, which is given in Chapter 4.

The present work is an extension of Giove (2002)'s work in the light of *Fuzzy Preference in interval order.* As DM's rationale and attitude are reflected in the latter index, the present work logically yields an improved solution.

8.2. Relevant preliminaries

8.2.1. *The original TOPSIS*

The Technique for Order Preference by Similarity to an Ideal Solution (TOPSIS) is presented in Chen & Hwang (1992), with reference to Hwang & Yoon (1981). The basic principle is that the chosen alternative should have the shortest distance from the ideal solution and the farthest distance from the negative-ideal solution.

The well-known TOPSIS procedure is summarized in the following steps:

1) *Calculate the normalized decision matrix.* The vector normalization is used for computing r_{ij}, which is given as

$$r_{ij} = x_{ij} \Big/ \sqrt{\sum_{i=1}^{m} x_{ij}^2}, \quad i = 1,\ldots\ldots,m; \quad j = 1,\ldots\ldots,n. \tag{8.2.1.1}$$

2) *Calculate the weighted normalized decision matrix.* The weighted normalized value v_{ij} is calculated as

$$v_{ij} = w_j r_{ij}, \quad i = 1,\ldots\ldots,m; \quad j = 1,\ldots\ldots,n, \tag{8.2.1.2}$$

where, w_j is the weight of the jth attribute or criterion and $\sum_{j=1}^{n} w_j = 1$.

3) *Identify the ideal and negative-ideal solution.* The A^* and A^- are defined in terms of the weighted normalized values as follows

$$\begin{aligned} A^* &= \left\{ v_1^*, v_2^*, \ldots, v_j^*, \ldots, v_n^* \right\} \\ &= \left\{ (\max_i v_{ij} \,|\, j = 1,\ldots.,n) \,\middle|\, i = 1, \ldots.., m \right\} \end{aligned} \tag{8.2.1.3}$$

and

$$\begin{aligned} A^- &= \left\{ v_1^-, v_2^-, \ldots, v_j^-, \ldots, v_n^- \right\} \\ &= \left\{ (\min_i v_{ij} \,|\, j = 1,\ldots.,n) \,\middle|\, i = 1, \ldots.., m \right\} \end{aligned} \tag{8.2.1.4}$$

4) *Calculate the separation measures.* The separation of each alternative from the ideal solution can be measured by using the n-dimensional Euclidean distance as follows

$$D_i^* = \sqrt{\sum_{j=1}^{n} (v_{ij} - v_j^*)^2}, \quad i = 1, \ldots\ldots, m. \tag{8.2.1.5}$$

Similarly, the separation from the negative-ideal solution is given as

$$D_i^- = \sqrt{\sum_{j=1}^{n} (v_{ij} - v_j^-)^2}, \quad i = 1, \ldots\ldots, m. \tag{8.2.1.6}$$

5) *Calculate the relative closeness to the ideal solution.* The relative closeness of the alternative A_i with respect to A^* is defined as

$$C_i^* = D_i^- \Big/ \left(D_i^* + D_i^-\right), \qquad i = 1, \ldots\ldots, m. \tag{8.2.1.7}$$

Note that $0 \le C_i^* \le 1$. Then $C_i^* = 1$ if $A_i = A^*$ and $C_i^* = 0$ if $A_i = A^-$.

6) *Rank the preference order.* Choose an alternative with the maximum C_i^* or rank alternatives according to C_i^* in descending order.

8.2.2. *A note on Giove (2002)'s Bag based Interval-TOPSIS*

Rebai (1993) introduced the BB-TOPSIS method (Bag based TOPSIS) in the context of an MCDM problem in which both numerical and logical data appear in the criteria definitions. Here we recall the main points only keeping our notations as far as consistent to the original one. A fuzzy bag $B(X)$ is defined over a finite set X, $card(X) = n$ as the following function $B_X : X \to \Re^{2n}$:

$$\begin{aligned} B_X &= \{C_1(X)/w_1,\ C_2(X)/w_2,\ \ldots\ \ldots,\ C_n(X)/w_n\} \\ &= \left\{\prod_{j=1}^{n} C_j(X)/w_j\right\}, \end{aligned} \tag{8.2.2.1}$$

where, X is an alternative described against the criterion $\{C_j(X)\}$, and w_j is the weight of the j-th criterion, while the symbol $\prod$ denotes the Cartesian product.

Rebai (1993) defined two bags B_X and B_Y over X and Y, with $card(X) = card(Y)$ as follows

$$\left.\begin{aligned} \textit{cardinality:} \quad & |B_X| = \sum_{j=1}^{n} C_j(X) w_j \\ \textit{symmetric difference:} \quad & B_X \Delta B_Y = \left\{ \prod_{j=1}^{n} \left|C_j(X) - C_j(Y)\right| w_j \right\} \\ \textit{distance:} \quad & d(B_1, B_2) = \left|B_X \Delta B_Y\right| \end{aligned}\right\} \tag{8.2.2.2}$$

Among other distance functions the above one was chosen as a suitable one (refer Rebai (1993) as well as Giove (2002) for detail). A bag can then be assigned to each alternative taking the values for each criterion and relative weights. Given an alternative, dominance value for a criterion defines a superiority (or inferiority or non-inferiority, depending on the nature of the criterion) score over other alternative values under the same criterion. For instance, the superiority score for the alternative X under j-th criterion is given by

$$S_j(X) = card\left\{ Y: \ C_j(Y) < C_j(X), \quad Y \in A \right\}, \tag{8.2.2.3}$$

where A is the set of alternatives. Then following bag defines the Multi-attribute Canonical Fuzzy Bag with the superiority score as

$$MCFB^{>}(X) = \left\{ \prod_{j=1}^{n} S_j(X) / C_j \right\}. \tag{8.2.2.4}$$

The ideal and anti-ideal alternatives are defined as

$$MCFB_{>}^{+}(X) = \vee MCFB^{>}(X) = \left\{ \prod_{j=1}^{n} \vee S_j(X) / C_j \right\}, \tag{8.2.2.5}$$

$$MCFB_{>}^{-}(X) = \wedge MCFB^{>}(X) = \left\{ \prod_{j=1}^{n} \wedge S_j(X) / C_j \right\}, \tag{8.2.2.6}$$

where, the symbols $\vee$ and $\wedge$ stand for the operators, maximum and minimum, respectively. Next for each of the given alternatives $X \in A$, the algorithm computes two sets of distances:

Set 1: distance between $MCFB^{>}(X)$ and the ideal bag $MCFB_{>}^{+}(X)$ and

Set 2: distance between $MCFB^{>}(X)$ and the anti-ideal bag $MCFB_{>}^{-}(X)$.

The two sets of distances are given by

$$d_{>}^{+} = d\left(MCFB^{>}(X),\ MCFB_{>}^{+}(X)\right) \tag{8.2.2.7}$$

$$d_{>}^{-} = d\left(MCFB^{>}(X),\ MCFB_{>}^{-}(X)\right) \tag{8.2.2.8}$$

and the separation index, which is similar to (8.2.1.7), is given as

$$C_{>}(X) = \frac{d_{>}^{-}}{d_{>}^{-} + d_{>}^{+}}. \tag{8.2.2.9}$$

Finally, the alternatives $X \in A$, are ranked according to the value of the index $C_{>}(X)$, as is done in the classical TOPSIS method.

Giove (2002) extended Rebai (1993)'s aforesaid BB-TOPSIS in the case of interval data. Interval analysis is a simple and intuitive way to treat data uncertainty for complex decision problems and is used in a lot of practical applications. Giove used $\mathcal{A}$-index for measuring 'how much an interval is greater than any other interval number' but in a slightly changed version (let us call it *Giove's modification*) as follows

$$\xi(X_1 > X_2) = \frac{m(X_1) - m(X_2)}{w(X_1) + w(X_2) + 1}, \tag{8.2.2.10}$$

where, X_1 and X_2 are any two interval numbers, and $m(X_1)$, $m(X_2)$, $w(X_1)$ and $w(X_2)$ are their means and widths respectively, with their as usual meaning. Here the index value can be positive or negative depending on whether $m(X_1)$ is greater than $m(X_2)$ or not. There is another interesting change in the denominator of the index; a value equal to +1 is added here in order to remove the condition, $w(X_1) + w(X_2) \neq 0$, as was stated in the original $\mathcal{A}$-index. In Giove (2002)'s opinion this has been done "to avoid loss of significance where both the spreads are null". However it has not been explained further whether the modification really improves the *comparability of the index* as a whole. (Chapter 3 of this book gives different properties and results of $\mathcal{A}$-index.) A discussion on the *Giove's modification* is given in Section 8.3.

According to Giove's proposal, for a set of interval data, the superiority score for the alternative $X \in \{X_i \mid i = 1, 2, ..., m\}$, where $X_i = X_{ij} = [(x_{ij})_L, (x_{ij})_R]$ under the j-th criterion, is given by

$$IS_j(X) = \sum_{\substack{i=1 \\ i \notin g(X)}}^{m} \xi_j(X > X_{ij}), \quad \text{for any } j = 1, 2, ..., n, \qquad (8.2.2.11)$$

where $g(X)$ is the index function for the set of alternatives $\{X_1, X_2, ..., X_m\}$ such that $g(X_i) = i$ and $IS_j(X)$ is defined as a replacement for the interval extension of $S_j(X)$ given in (8.2.2.3).

In Giove (2002)'s paper, the description of rest of the procedure is almost similar to that of Rebai (1993)'s BB-TOPSIS. Expressions (8.2.2.4) to (8.2.2.9) are rewritten as expressions (8.2.2.12) to (8.2.2.17) with changes in the variable names suitable for extension of BB-TOPSIS with interval data.

The interval-valued Multi-attribute Canonical Fuzzy Bag with the superiority score is defined as

$$IMCFB^{>}(X) = \left\{ \prod_{j=1}^{n} IS_j(X) / w_j \right\}. \qquad (8.2.2.12)$$

Subsequently, the ideal and anti-ideal bags are defined as

$$IMCFB_{>}^{+}(X) = \vee\, IMCFB^{>}(X) = \left\{ \prod_{j=1}^{n} \vee IS_j(X) / w_j \right\}, \qquad (8.2.2.13)$$

$$IMCFB_{>}^{-}(X) = \wedge\, IMCFB^{>}(X) = \left\{ \prod_{j=1}^{n} \wedge IS_j(X) / w_j \right\}. \qquad (8.2.2.14)$$

The two sets of distances are given by

$$id_{>}^{+} = d\left(IMCFB^{>}(X),\ IMCFB_{>}^{+}(X) \right) \qquad (8.2.2.15)$$

$$id_{>}^{-} = d\left(IMCFB^{>}(X),\ IMCFB_{>}^{-}(X) \right) \qquad (8.2.2.16)$$

and the separation index as

$$IC_{>}(X) = \frac{id_{>}^{-}}{id_{>}^{-} + id_{>}^{+}}. \tag{8.2.2.17}$$

Finally, the alternatives $X \in A$, are ranked according to the value of the index $IC_{>}(X)$.

8.3. Fuzzy Preference Ordering of Interval Attributes in I-TOPSIS

Giove (2002)'s modification in the basic structure of $\mathscr{A}$-index as given in (8.2.2.10) has the following two limitations:

(i) The main objective of $\mathscr{A}$-index is to rank any two interval-numbers with the help of a DM's objective preference. As a value judgment index it performs efficiently and reveals thorough consistency to our intuition. Findings of this index are also not contradictory to the existing main approaches in the literature (for example, Ishibuchi & Tanaka (1990)'s preference relations and Kundu (1997)'s fuzzy leftness relation for interval ranking). Moreover it has shown some advantages over the other approaches.

Whereas Giove (2002)'s modified $\mathscr{A}$-index can neither explain nor set apart the basic cases like definitely acceptable or possibly acceptable situation of 'one interval is superior (or inferior) than the other'. For example, let B_1, B_2 and B_3 be three intervals such that

$$B_1 = [1,\ 2] = \langle 1.5,\ .5 \rangle,$$

$$B_2 = [2,\ 3] = \langle 2.5,\ .5 \rangle \text{ and}$$

$$B_3 = [2,\ 10] = \langle 6,\ 4 \rangle,$$

then using the modified index (8.2.2.10), we have,

$$\begin{aligned} \xi(B_2 > B_1) &= .5 \\ \xi(B_3 > B_1) &= .82 \end{aligned} \tag{8.3.1}$$

Here each of the intervals B_2 and B_3 are weakly mutually exclusive to a same interval B_1. But Giove's modification does not show any resemblance between the comparing pairs. Whereas $\mathscr{A}$-index shows that

$$\mathscr{A}(B_1 \odot B_2) = 1$$
$$\mathscr{A}(B_1 \odot B_3) = 1 \qquad (8.3.2)$$

$\mathscr{A}$-index interprets that both of intervals B_2 and B_3 are equally and definitely preferred to interval B_1 as the maximizing alternatives. Can the modified index be explained in a better way unless it is treated as some sort of 'difference between the intervals'?

(ii) If the modified index is treated as some sort of 'difference between the intervals', then the incompatibility of the following kind can arise.

If the intervals $B_1 = [1,\ 2] = <1.5,\ .5>$ and $B_2 = [2,\ 3] = <2.5,\ .5>$ are multiplied by scalars $k = \{1, 5, 10\}$, then we have $(B_{1^k},\ B_{2^k}), \forall k$, as the comparing pairs of intervals, where, $B_{t^k} = k.B_t$, $t = \{1,2\}$, (for example, for $t = 2, k = 10$, B_{t^k} is $B_{2^{10}} = 10.B_2 = [20,\ 30] = <25,\ 5>$),

We get the following results for the comparing pairs of intervals:

$$B_{2^1} - B_{1^1} = [0,\ 2] = \ <1,\ 1>$$
$$B_{2^5} - B_{1^5} = [0,\ 10] = \ <5,\ 5>$$
$$B_{2^{10}} - B_{1^{10}} = [0,\ 20] = \ <10,\ 10>$$

$$\xi(B_{2^1} > B_{1^1}) = .5 \qquad \mathscr{A}(B_{1^1} \odot B_{2^1}) = 1$$
$$\xi(B_{2^5} > B_{1^5}) = .833 \qquad \mathscr{A}(B_{1^5} \odot B_{2^5}) = 1$$
$$\xi(B_{2^{10}} > B_{1^{10}}) = .9 \qquad \mathscr{A}(B_{1^{10}} \odot B_{2^{10}}) = 1$$

This example clearly shows that $\mathscr{A}$-index works more consistently than the Giove's modification in defining DM's objective acceptability of whether an interval is superior or inferior to the other on the basis of their mean values and the levels of uncertainty.

Moreover neither of the indices is suitable for being used as the concept of *difference between the intervals* as we see in real analysis. In Section 3.2, the definition of $\mathscr{A}$-index clearly states that it does not work for comparing two real numbers. We see same sort of limitation for defining extended division between two intervals (refer Section 1.4).

In another argument, interpretation of $\mathscr{A}$-index does not conform with the plausibility of defining *the grade of acceptability of superiority of a real number to another real number*. At the end of Section 3.2, as an important point it is noted that the concept of $\mathscr{A}$-index as a ranking function is in no way be analogous to the concept of *difference between the intervals*, as we see in real analysis.

However, as a complete preference index, limitations of $\mathscr{A}$-index are discussed in Section 4.2. There we have introduced our 2nd index, the *fuzzy preference ordering* scheme for comparing any two interval numbers as an advanced scheme. Reflection of a pessimistic or an optimistic DM's attitude in making a decision is a salient feature of this index (Section 4.6 presents a sequence of arguments in detail).

The procedure for ranking many intervals simultaneously is given and used elaborately in Chapter 5. Similar sort of approach is applied here in I-TOPSIS to get better version of normalization in the processing of interval data considering (8.2.2.3) and (8.2.2.11) in the procedures stated above.

For a maximization problem, any pair of intervals A and B are ordered as (A, B) if $m(A) \le m(B)$ and are classified into two sets, S_1 and S_2 as follows:

$$\begin{aligned} &\text{(i)} \quad (A,\ B) \in S_1,\ \text{if}\ \ w(A) \ge w(B), \\ &\text{(ii)} \quad (A,\ B) \in S_2,\ \text{if}\ \ w(A) < w(B). \end{aligned} \tag{8.3.3}$$

Then,

- For $(A,\ B) \in S_1$ unless A and B are identical, B *is always the best choice.*
- For $(A,\ B) \in S_2$, a membership function of *fuzzy preferences between A and* B can be constructed.

Here we try to reiterate the findings of Chapter 4 in brief in order to frame a computational routine for ranking any two *utility intervals* from among a set of n. Ranking of utility intervals by-default assumes a set of assumptions only for a maximization problem setup.

We define a fuzzy set B' as the *rejection of B* compared to A for any pair of utility intervals (A, B), where, $B' = \{(A,B) \in S_2 \mid m(A) \le m(B), w(A) < w(B)\}$, then

$$\pi_{B'}(A,B) = \max \left\{0, \frac{m(A) - (b_L + w(A))}{m(B) - (b_L + w(A))}\right\}^p \tag{8.3.4}$$

where, $\pi_{B'}(A,B) \in [0,\ 1]$ and the parameter p is the (non-fuzzy) level of pessimism for any decision maker and $p \in [(1/M),\ M]$, where M is a finite large number. If $p = 1/M$, then the decision maker is said to be of an absolutely pessimistic person who practically does not take any risk. Similarly, for an absolutely optimistic DM, we use $p = M$. The level of pessimism, at $p = 1$, may be regarded as the reflection of a moderate DM's attitude against the aforementioned extreme odds. (Refer Sections 4.3 – 4.4 for detail.)

We define another fuzzy set, A', *rejection of A* for $(A, B) \in S_2$, as the complement of B' as

$$\pi_{A'}(A, B) = 1 - \pi_{B'}(A, B).$$

We can interpret different values of $\pi_{B'}$ as follows:

if $\pi_{B'} = 1$, then B is definitely rejected and A is definitely accepted;

if $\pi_{B'} = 0$, then B is definitely accepted and A is definitely rejected and

if $\pi_{B'}$ is in between 0 and 1; then there exists different degrees of *rejection of B*.

Hence we can frame a computational routine as

1. if $A \equiv B$, then put $\pi_{A'}(A, A) = .5$. (8.3.5)
2. if $(A, B) \in S_1$, then put $\pi_{B'}(A, B) = 1 - \pi_{A'}(A, B) = 0$. (8.3.6)
3. if $(A, B) \in S_2$, then (taking $p = 1$) compute

$$\pi_{B'}(A, B) = 1 - \pi_{A'}(A, B) = \max\left\{0, \frac{m(A) - (b_L + w(A))}{m(B) - (b_L + w(A))}\right\}. \tag{8.3.7}$$

Therefore, for a set of interval data from the MCDM problem (8.1.1), the *superiority score* for the alternative $X \in \{X_i = [(x_i)_L, (x_i)_R] \,/\, i = 1, 2, ..., m\}$ under the j-th criterion is defined for all $j = 1, 2, \cdots, n$, as follows

$$IS_j(X_i) = \sum_{k=1}^{m} \pi_i^j(X_i, X_k), \quad k \in \{i = 1, 2, \cdots, m\}; \tag{8.3.8}$$

where, $IS_j(X_i)$ is obtained on the basis of fuzzy preference ordering for intervals and it replaces Giove (2002)'s $IS_j(X)$ given in (8.2.2.11).

Subsequently, the Interval-valued Multi-attribute Canonical Fuzzy Bag with the superiority score given in (8.2.2.12) is redefined as

$$IMCFB^{>}(X) = \left\{\prod_{j=1}^{n} IS_j(X_i)/w_j\right\}. \tag{8.3.9}$$

Following (8.2.2.13) and (8.2.2.14), the ideal and anti-ideal bags are then redefined as

$$IMCFB_{>}^{+}(X) = \vee IMCFB^{>}(X)$$

$$= \left\{ \left(\operatorname*{Max}_{i=1}^{m} IS_1(X_i) \Big/ w_1 \right), \left(\operatorname*{Max}_{i=1}^{m} IS_2(X_i) \Big/ w_2 \right), \cdots, \left(\operatorname*{Max}_{i=1}^{m} IS_n(X_i) \Big/ w_n \right) \right\}$$

$$= \left\{ \prod_{j=1}^{n} \left(\operatorname*{Max}_{i=1}^{m} IS_j(X_i) \Big/ w_j \right) \right\} \tag{8.3.10}$$

$$IMCFB_{>}^{-}(X) = \wedge IMCFB^{>}(X)$$

$$= \left\{ \left(\operatorname*{Min}_{i=1}^{m} IS_1(X_i) \Big/ w_1 \right), \left(\operatorname*{Min}_{i=1}^{m} IS_2(X_i) \Big/ w_2 \right), \cdots, \left(\operatorname*{Min}_{i=1}^{m} IS_n(X_i) \Big/ w_n \right) \right\}$$

$$= \left\{ \prod_{j=1}^{n} \left(\operatorname*{Min}_{i=1}^{m} IS_j(X_i) \Big/ w_j \right) \right\} \tag{8.3.11}$$

The two sets of distances and the separation index are given as follows:

$$id^{+}(X_i) = d\left(IMCFB^{>}(X),\ IMCFB_{>}^{+}(X) \right)$$

$$= \sum_{j=1}^{n} \left(\operatorname*{Max}_{i=1}^{m} IS_j(X_i) - IS_j(X_i) \right) w_j \tag{8.3.12}$$

$$id^{-}(X_i) = d\left(IMCFB^{>}(X),\ IMCFB_{>}^{-}(X) \right)$$

$$= \sum_{j=1}^{n} \left| \left(\operatorname*{Min}_{i=1}^{m} IS_j(X_i) - IS_j(X_i) \right) \right| w_j \tag{8.3.13}$$

$$\zeta_i = \frac{id^{-}(X_i)}{id^{-}(X_i) + id^{+}(X_i)}. \tag{8.3.14}$$

Finally, the alternatives $X_i \in A$ are ranked according to the index value ζ_i.

Alternatives	Maximizing Criteria with weights		
	$C_1/0.3$	$C_2/0.5$	$C_3/0.2$
X_1	[5,7]	[4,8]	[1,9]
X_2	[4,8]	[2,6]	[5,7]
X_3	[6,10]	[3,7]	[2,6]

Table 8.1: The Interval MCDM problem

8.4. A numerical example

Here we cite an Interval Multi-criteria Decision Problem which was used earlier by Giove (2002).

We have given three sets of interval data on three criteria C_1, C_2 and C_3 with given weights 0.3, 0.5 and 0.2 respectively. The problem is to rank the alternatives X_1, X_2 and X_3. The problem data is given in Table 8.1.

Table 8.2 denotes pair-wise fuzzy rejection grades for each $C_j, \forall j = 1,2,3,$ where

$$\pi_{i'}^{j}\{X_i, X_k\},\ k \in \{i\}$$

represents rejection of X_i over X_k for the jth criterion. The pair of intervals $\{X_i, X_k\}$ is an unordered pair according to (8.3.3) and therefore made ordered pairs before applying the rules (8.3.5) – (8.3.7) for getting the superiority scores.

Table 8.3 gives the superiority Score for alternative X_i for the jth criterion

$$IS_j(X_i) = \sum_{k=1}^{3} \pi_{i'}^{j}(X_i, X_k), \quad k \in \{i = 1,2,3\}; \quad \text{for any } j = 1,2,3,$$

Now, for the jth criterion, $\forall j$, the column maximum, $S_j^{\max}$ and the column minimum, $S_j^{\min}$ are computed as follows

$$S_j^{\max} = \operatorname*{Max}_{i=1}^{3} IS_j(X_i) \qquad\qquad S_j^{\min} = \operatorname*{Min}_{i=1}^{3} IS_j(X_i)$$

Alternatives	C_1	C_2	C_3
X_1	$\pi_{1'}^{1}\{X_1, X_1\} = .5$ $\pi_{1'}^{1}\{X_1, X_2\} = 0$ $\pi_{1'}^{1}\{X_1, X_3\} = 1$	$\pi_{1'}^{2}\{X_1, X_1\} = .5$ $\pi_{1'}^{2}\{X_1, X_2\} = 0$ $\pi_{1'}^{2}\{X_1, X_3\} = 0$	$\pi_{1'}^{3}\{X_1, X_1\} = .5$ $\pi_{1'}^{3}\{X_1, X_2\} = 1$ $\pi_{1'}^{3}\{X_1, X_3\} = .5$
X_2	$\pi_{2'}^{1}\{X_2, X_1\} = 1$ $\pi_{2'}^{1}\{X_2, X_2\} = .5$ $\pi_{2'}^{1}\{X_2, X_3\} = 1$	$\pi_{2'}^{2}\{X_2, X_1\} = 1$ $\pi_{2'}^{2}\{X_2, X_2\} = .5$ $\pi_{2'}^{2}\{X_2, X_3\} = 1$	$\pi_{2'}^{3}\{X_2, X_1\} = 0$ $\pi_{2'}^{3}\{X_2, X_2\} = .5$ $\pi_{2'}^{3}\{X_2, X_3\} = 0$
X_3	$\pi_{3'}^{1}\{X_3, X_1\} = 0$ $\pi_{3'}^{1}\{X_3, X_2\} = 0$ $\pi_{3'}^{1}\{X_3, X_3\} = .5$	$\pi_{3'}^{2}\{X_3, X_1\} = 1$ $\pi_{3'}^{2}\{X_3, X_2\} = 0$ $\pi_{3'}^{2}\{X_3, X_3\} = .5$	$\pi_{3'}^{3}\{X_3, X_1\} = .5$ $\pi_{3'}^{3}\{X_3, X_2\} = 1$ $\pi_{3'}^{3}\{X_3, X_3\} = .5$

Table 8.2: The normalization operation: Computation of pair-wise superiority scores for each criterion.

Alternatives	$IS_1(X_i)$	$IS_2(X_i)$	$IS_3(X_i)$
X_1	1.5	.5 → Col. Min	2 → Col. Max
X_2	2.5 → Col. Max	2.5 → Col. Max	.5 → Col. Min
X_3	.5 → Col. Min	1.5	2

Table 8.3: The superiority Score for alternative X_i with reference to *j*-th criterion.

Alternatives	$id^{+}(X_i)$	$id^{-}(X_i)$
X_1	1.3	.6
X_2	.3	1.6
X_3	1.1	.8

Table 8.4: The TOPSIS distances

Next, according to (8.3.12) and (8.3.13), the TOPSIS distances i.e., the distances between the ideal solution set and the superiority score set for each X_i and the distances between the anti-ideal solution set and the superiority score set for each X_i, are computed as follows and given next in Table 8.4:

$$id^{+}(X_i) = \sum_{j=1}^{3} \left(S_j^{\max} - IS_j(X_i)\right).w_j$$

$$id^{-}(X_i) = \sum_{j=1}^{3} \left|S_j^{\min} - IS_j(X_i)\right|.w_j$$

Now, through the *separation index*,

$$\zeta_i = \frac{id^{-}(X_i)}{id^{+}(X_i) + id^{-}(X_i)},$$

we have the following rank of alternatives given in Table 8.5, where rank I is assigned to the alternative with highest preference, rank II gives the alternative with the next maximum preference and rank III the lowest.

Alternatives	ζ_i	Rank
X_1	.3158	III
X_2	.8421	I
X_3	.4210	II

Table 8.5: Separation index for ranking

8.5. Conclusion

This chapter presents a modified version of Giove (2002)'s I-TOPSIS approach for interval-valued MCDM problems. The modification is done using *Fuzzy Preference in interval order* instead of applying $\mathscr{A}$-index for comparing intervals. Obviously our result is different from Giove's result. A numerical comparison, however, is not practically possible. As DM's rationale and preference attitude is reflected in the fuzzy preference-ordering scheme, the present algorithm logically yields an improved solution.

8.6. References

M.A. **Abo-Sinna** & A.H. **Amer (2005)**, Extensions of TOPSIS for multi-objective large-scale nonlinear programming problems, *Applied Mathematics and Computation* 162 (1): 243–256.

R.E. **Bellman** & L.A. **Zadeh (1970)**, Decision-making in a fuzzy environment, *Management Science* 17 (4): 141 – 164.

S.-J. **Chen**, C.-L. Hwang & F. P. Hwang **(1992)**, *Fuzzy multiple attribute decision making*, Springer-Verlag, Berlin.

C.T. **Chen (2000)**, Extensions of the TOPSIS for group decision-making under fuzzy environment, *Fuzzy Sets and Systems* 114 (1): 1 – 9.

C.T. **Chen**, C.T. Lin & S.F. Huang **(2006)**, A fuzzy approach for supplier evaluation and selection in supply chain management, *International Journal of Production Economics* 102 (2): 289 – 301.

M. **Delgado**, J.L. Verdegay, M.A. Vila **(1989)**, A general model for fuzzy linear programming, *Fuzzy Sets and Systems* 29 (1): 21–29.

J.S. **Dyer**, P.C. Fishburn, R.E. Steuer, J. Wallenius & S. Zionts **(1992)**, Multiple criteria decision making and multiattribute utility theory: The next ten years, *Management sciences* 38 (5): 645-654.

S. **Giove (2002)**, Interval TOPSIS for Multicriteria Decision Making, in: M. Marinaro & R. Tagliaferri (Eds.): WIRN VIETRI 2002, *Lecture Notes in Computer Science* 2486, Springer-Verlag Berlin Heidelberg: 56-63.

F. **Herrera**, E. Herrera-Viedma & J.L. Verdegay **(1996)**, A model of consensus in group decision-making under linguistic assessments, *Fuzzy Sets and Systems* 78 (1): 73 – 87.

H.M. **Hsu** & C.T. **Chen (1996)**, Aggregation of fuzzy opinions under group decision-making, *Fuzzy Sets and Systems* 79 (3): 279 – 285.

C.L. **Hwang** & K. **Yoon (1981)**, Multiple Attribute Decision Making, *Lecture Notes in Economics and Mathematical Systems*, 186, Springer-Verlag, Berlin.

A. **Rebai (1993)**, BB-TOPSIS: a bag based technique for order preference by similarity to ideal solution, *Fuzzy Sets and Systems* 60: 143 – 162.

J. **Teghem Jr.**, C. Delhaye & P.L. Kunsch **(1989)**, An interactive decision support system (IDSS) for multicriteria decision aid, *Mathematical and Computer Modelling* 12 (10/11): 1311 – 1320.

Chapter 9

Concluding Remarks and the Future Scope

9.1. Introduction

The use of interval parameters is possibly the simplest but yet an intuitive way to introduce data uncertainty for large and complex decision problems. The way of argument as well as the advantages in handling data uncertainty in the form of interval numbers has attracted many researchers. Consequently, the problem of interval number ordering has become a matter of perennial interest because of its direct relevance to the practical modelling and optimization in the real world processes under imprecise and uncertain environment. However, one main dilemma in using interval data for decision problems was perhaps in making the choice of an appropriate interval order relation. Earlier it was believed and was often argued that crisp intervals, theoretically, could only be partially ordered and hence all were not comparable. Nevertheless, when interval numbers are started to be aptly used as uncertain parameters in practical applications, the question of choice among the interval alternatives is started brewing. This unattended question of ranking any two interval numbers has created a new enthusiasm among the researchers during only the last two decades.

A. Sengupta and T.K. Pal: Fuzzy Preference Ordering, STUDFUZZ 238, pp. 155–159.
springerlink.com

In this research monograph primarily we have tried to congregate as many as relevant works on the ranking of interval numbers. We have discussed and compared different aspects of thirteen prominently different ranking indices published so far. (Obviously there are a few publications, which may have been unknowingly left out of this discussion. Non availability of timely resources might have been another important reason. One or two works, however, may be considered intentionally omitted mainly due to their indistinguishability from the other in concept and/or form.) Next we have tried to study the efficiency and applicability of these ranking alternatives in realistic decision problems. Various applications of interval numbers as an imprecise and uncertain data source in decision problems, as a whole, may be considered as one key intention underneath this book.

9.2. Chapter Summary and Conclusion

Chapter 1 is the introductory chapter of this book. Apart from sketching the motivation of this book, this chapter gives an overview of representing uncertainty and imprecision in mathematical modeling. Genesis, notation and symbolism of interval numbers are discussed in detail.

Chapter 2 contains the first part of a comprehensive study on the existing ideas regarding comparison of interval numbers and identifies the lacunae that motivate research in this field.

In chapter 3, we have defined *Acceptability Index*, or in short, $\mathscr{A}$-index, for comparing any two intervals through DM's satisfaction. Various properties of this index have also been given. However, this chapter mainly approaches to define a satisfactory crisp equivalent system of an interval inequality constraint as an application of $\mathscr{A}$-index. Based on the interpretation of interval inequality constraints and interval-valued objective function, an interval linear programming problem is solved.

Chapter 4 presents a brief discussion on the strength and limitation of $\mathscr{A}$-index over the other approaches. $\mathscr{A}$-index is described here as a value judgment index for comparing any two real intervals. As a preference index, this index has also been called *an optimistic decision-maker's preference index*. Next, in order to discuss a *pessimistic* or *risk-averse* DM's point of view, *fuzzy preference ordering* between any two interval-attributes (say, profit, cost of production etc.) has been introduced in this chapter. As a *generalized preference index*, different DMs' choices with different degrees of pessimism have also been considered and explained with suitable numerical example. Efficiency of these two indices over the other ranking strategies has been summed up in two interesting tabular set-ups. In continuation to the discussion initiated in Chapter 2, the last section of Chapter 4 studies some recent ranking indices too.

Chapter 5 presents a solution procedure of a shortest path problem of which the networking arcs are represented as interval numbers. The impersonal as well as prejudiced fuzzy preference ordering between intervals defined in chapter 4 has been extended here for ranking a set of intervals simultaneously and executed in a network problem. With this ranking strategy, Dijkstra's algorithm has been generalized and made simple and effective for finding permanent nodes from among the temporary nodes at every stages of iteration. Though imprecision is embedded in the problem definition, the solution procedure is well capable of providing an impersonal solution as well as many prejudiced solutions on the basis of subjective interaction only (and not by any objective query) with the human decision maker. Another example of a large problem is illustrated in this chapter.

Chapter 6 describes a Travelling Salesman Problem using interval arcs. The salesman here wants to visit a certain number of cities allotted to him, where the *cost* or *distance* or *time* of journey between every pair of cities are considered as interval numbers as these may fluctuate considerably depending on traffic condition, payload etc.

The objective of this work in a narrower sense is to select such a route that starts from the salesman's home city, passes through each city allotted to him once and only once, and returns to his home city at lowest possible cost (or in least time or in the shortest possible distance). In this problem, we define each arc in terms of real intervals. However, the objective in wider sense was to find another application of interval parameters in decision problems and to define a specific choice between existing interval order relations. Here, both of the value-based ranking index and the preference ranking index for interval attributes have been used here. For finding SIVs, the acceptability index is used and for selecting the best interval path from the zero-centered interval optimal paths a preference ranking methodology is used. In this chapter we have considered a simple ITSP. In complex decision situation, when optimal and near-optimal solutions are in plenty, then fuzzy preference ordering with its totality may be regarded the most suitable one.

The aim of chapter 7 is to define an Interval-valued Transportation Problem with Multiple Penalty Factors (ITPMPF) and describe a procedure for getting a satisfactory compromise solution. In any inexact programming problem, the first task is to explain the *meaning* of the objective defined in the said environment. So, in the light of *interval order relations* discussed earlier in Chapters 2 – 4, maximization / minimization of an interval objective function is interpreted and finally a composite objective function of an ITPMPF is defined. The ITPMPF is reduced into a standard LPP structure and solved. One noticeable feature of this work is that it reflects DM's pessimistic or optimistic bias in achieving the compromise solution. Another important aspect of this approach is the flexibility with which the DM finds more hold into the problem than any other approach.

Giove's I-TOPSIS (Interval TOPSIS) used the modified acceptability index. Chapter 8 brings an extension of the I-TOPSIS in the light of an improved work Fuzzy Preference in interval order.

9.3. Future Research Agenda

Here we would like to sketch or just can only point out some of the possible avenues of the future research.

- In continuation to the discussion in Chapter 2 – 4, all interval and fuzzy ranking strategies may be compared and substantiated case to case basis.
- An interactive approach to solve a Multiobjective interval linear programming problem: with an interactive approach a DM learns gradually more and more about the problem and his feedback is used to guide the solution of the problem to a more favourable one. Thus a generalized and more credible procedure for solving a class of interval linear programming problems (and also some nonlinear programming problems) may be formulated.
- Duality conditions in interval optimization problem may be studied further and validated with real considerations.
- As a preference index, fuzzy preference ordering of interval attributes has been defined from pessimistic and optimistic DMs' point view. Larger and more useful application area may be found in the fields of Market study and research.
- Extension of VIKOR method for decision making problem with interval data using fuzzy preference ordering scheme.
- The aggregation of preference ranking has wide applications in social choice, vendor selection, voting system, etc. Association of interval data with aggregated preference ranking in group decision making situation requires considerable attention.
- In spite of some direct importance of the Interval Travelling Salesman Problem, quite a lot of practical applications of discrete optimization can be put in this form. So, in future, development of algorithm for large and complex problems is necessary.
- Methods of grey related analysis to multiple attribute decision making problems with interval numbers may be investigated.
- In comparison to fuzzy preference ordering of interval utilities, the efficiency and effectiveness of interval comparison based on Dempster-Shafer Theory of Evidence may be investigated.
- A more generalized decision-support-system using interval data may be thought of to promote decision in an inexact environment.

In fact, all of these works may inspire and enlighten some interesting avenues to further research both in theory and application using interval parameters. Researchers may find important research applications particularly in the areas of soft sciences like Social Science, Economics, Management studies, etc., where the citations of incomplete/uncertain/imprecise data sources are in plenty, but more and more application of interval arithmetic and ranking methodologies are yet to find enough attention.

Index

List of Figures

List of Tables

Kolibri 360 GmbH
Ferdinand-Jühlke-Straße 7
99095 Erfurt, Deutschland
produktsicherheit@kolibri360.de